Data Collection in Fragile States

Johannes Hoogeveen · Utz Pape
Editors

Data Collection in Fragile States

Innovations from Africa and Beyond

Editors
Johannes Hoogeveen
World Bank
Washington, DC, USA

Utz Pape
World Bank
Washington, DC, USA

ISBN 978-3-030-25119-2   ISBN 978-3-030-25120-8 (eBook)
https://doi.org/10.1007/978-3-030-25120-8

© International Bank for Reconstruction and Development/The World Bank 2020. This book is an open access publication.

The opinions expressed in this publication are those of the authors/editors and do not necessarily reflect the views of the International Bank for Reconstruction and Development/The World Bank, its Board of Directors, or the countries they represent.

Open Access This book is licensed under the terms of the Creative Commons Attribution 3.0 IGO License (https://creativecommons.org/licenses/by/3.0/igo/), which permits use, sharing, adaptation, distribution and reproduction in any medium or format, as long as you give appropriate credit to the International Bank for Reconstruction and Development/The World Bank, provide a link to the Creative Commons licence and indicate if changes were made.

The use of the International Bank for Reconstruction and Development/The World Bank's name, and the use of the International Bank for Reconstruction and Development/The World Bank's logo, shall be subject to a separate written licence agreement between the International Bank for Reconstruction and Development/The World Bank and the user and is not authorized as part of this CC-IGO licence. Note that the link provided above includes additional terms and conditions of the licence.

The images or other third party material in this book are included in the book's Creative Commons licence, unless indicated otherwise in a credit line to the material. If material is not included in the book's Creative Commons licence and your intended use is not permitted by statutory regulation or exceeds the permitted use, you will need to obtain permission directly from the copyright holder.

The use of general descriptive names, registered names, trademarks, service marks, etc. in this publication does not imply, even in the absence of a specific statement, that such names are exempt from the relevant protective laws and regulations and therefore free for general use.

The publisher, the authors and the editors are safe to assume that the advice and information in this book are believed to be true and accurate at the date of publication. Neither the publisher nor the authors or the editors give a warranty, expressed or implied, with respect to the material contained herein or for any errors or omissions that may have been made. The publisher remains neutral with regard to jurisdictional claims in published maps and institutional affiliations.

This Palgrave Macmillan imprint is published by the registered company Springer Nature Switzerland AG
The registered company address is: Gewerbestrasse 11, 6330 Cham, Switzerland

Foreword

The world is becoming less safe and peaceful. According to the 2018 Global Peace Index prepared by the Institute for Economics and Peace, 42 countries experienced an increase in the intensity of internal conflict over the past decade, twice the number of countries that have improved. While progress is being made in certain areas—military spending declined slightly, for instance—peacefulness deteriorated as the intensity of conflict worsened.

Conflict has major costs, in terms of lives prematurely ended, human suffering and forgone development and economic opportunities. A civil war costs a medium-sized developing country the equivalent of 30 years of GDP growth; it takes 20 years for its trade levels to return to pre-war levels. To mitigate the long-term consequences of conflict on growth and poverty reduction, the World Bank Group is paying increasing attention to countries affected by conflict and violence. Since 2017, the World Bank Group has doubled its financial support for countries facing current or rising risks of fragility, opened special windows for assistance to refugees and host communities, and developed new financial instruments to support crisis preparedness and response.

v

vi   Foreword

For financing to be effective, a good understanding of the situation is essential. Without timely and reliable data, development interventions risk being based on anecdotal evidence, with all the risks that come with inadequate planning, poor designs, and ineffective targeting. Quality data are critical for development interventions to be effective but are hard to obtain in situations of violence and conflict. Worse, collecting good data is rarely a priority in situations where urgency trumps being deliberate.

This book offers a welcome reprieve from this habit. The authors care about collecting statistical information and have gone to great lengths to compile data in some of the world's most challenging circumstances. That they succeeded speaks to their tenacity and ability to think outside the box. The variety of approaches and solutions discussed means that many practitioners will find something of value in "Data Collection in Fragile Situations." The book effectively eliminates the notion that data cannot be collected in certain difficult circumstances. In doing so, it shifts the paradigm from "there are no data" to "how do we go about collecting data here?"

The innovations presented in this book are relevant beyond fragile situations, and the Poverty and Equity Global Practice I lead has started to apply approaches discussed here in other contexts. We are exploring the use of mobile phone surveys and permanent enumerators to strengthen statistical data collection for remote locations, many of which are small island states threatened by climate change. We are testing approaches to ask sensitive questions, for example to obtain better information about the occurrence of gender-based violence in World Bank projects. More generally, the innovations described in this book allow us to be more imaginative in creating feedback loops and introducing systematic learning in the World Bank's portfolio of projects.

These are just some of the ways in which the Poverty and Equity Global Practice is internalizing the innovations presented in this book. I am convinced that others too will find inspiration here. For readers who would like to know more, I urge them to contact the authors of the chapters directly. They will be more than happy to offer additional details or assistance. Contact details for all authors can be found in the contributor section.

Washington, DC, USA

Carolina Sánchez-Páramo
Senior Director, Poverty
and Equity Global Practice

Acknowledgements

This book benefited from the generous support of the Belgian TF 0A2158, the SPF TF, as well as the support of the management of the Poverty and Equity Global Practice. Feedback from participants at the 2018 Fragility Forum convinced us of the interest in this book. Hannah McNeish edited the document. The very constructive feedback from Paul Bance, Kathleen Beegle, Bernard Harborne, and Christina Wieser is gratefully acknowledged.

Contents

1 Fragility and Innovations in Data Collection   1
Johannes Hoogeveen and Utz Pape

Part I Innovations in Data Collection

2 Monitoring the Ebola Crisis Using Mobile Phone Surveys   15
Alvin Etang and Kristen Himelein

3 Rapid Emergency Response Survey   33
Utz Pape

4 Tracking Displaced People in Mali   51
Alvin Etang and Johannes Hoogeveen

5 Resident Enumerators for Continuous Monitoring   63
Andre-Marie Taptué and Johannes Hoogeveen

xii   Contents

6   A Local Development Index for the CAR and Mali   83
*Mohamed Coulibaly, Johannes Hoogeveen, Roy Katayama
and Gervais Chamberlin Yama*

Part II   Methodological Innovations

7   Methods of Geo-Spatial Sampling   103
Stephanie Eckman and Kristen Himelein

8   Sampling for Representative Surveys of Displaced
     Populations   129
*Ana Aguilera, Nandini Krishnan, Juan Muñoz,
Flavio Russo Riva, Dhiraj Sharma and Tara Vishwanath*

9   Rapid Consumption Surveys   153
Utz Pape and Johan Mistiaen

10   Studying Sensitive Topics in Fragile Contexts   173
Mohammad Isaqzadeh, Saad Gulzar and Jacob Shapiro

11   Eliciting Accurate Consumption Responses from
     Vulnerable Populations   193
Lennart Kaplan, Utz Pape and James Walsh

Part III   Other Innovations

12   Using Video Testimonials to Give a Voice to the Poor   209
Utz Pape

13   Iterative Beneficiary Monitoring of Donor Projects   215
Johannes Hoogeveen and Andre-Marie Taptué

Contents xiii

14 Concluding Remarks: Data Collection in FCV Environments 235
Johannes Hoogeveen and Utz Pape

Index 241

Notes on Contributors

Ana Aguilera works as an Urban Development Specialist in the Latin America and Caribbean Region. Her work focuses on improving city management with an emphasis on urban economics and spatial development. Her work also comprises survey management and design to measure living standards and socioeconomic indicators in countries such as Tanzania, South Africa, Sierra Leone, Lebanon, Jordan and the Kurdistan region. Ana has contributed to various World Bank's Urbanization Reviews, including Ethiopia, Nigeria, Turkey and Central America. In 2014 she was awarded with the Youth Innovation Fund for her work using Big Data to understand mobility patterns in cities. Prior to the World Bank, Ana worked as an Economist at CAF's Direction of Public Policy and Competitiveness. She has also worked in the public and private sectors in Latin America and the United States, as well as on policy consulting advising local and regional governments. Ana graduated as an Economist from Universidad Católica Andrés Bello in Caracas, and holds a M.Sc. in Public Policy from The University of Chicago.

Mohamed Coulibaly is a consultant in the Poverty and Equity Global Practice of the World Bank. He graduated from the National School of Statistics and Applied Economics (ENSEA) as an engineer in

xvi   Notes on Contributors

statistics and economics, he started his career at Bloomfield Investment Corporation gaining experience on country and sector risk assessment and public debt rating. Before joining the World Bank, Mohamed was a research officer at the Cabinet of the Minister of Planning and Development in Cote d'Ivoire. His current work focuses on evaluating local development, harmonizing household survey data in Sub-Saharan Africa, and assessing fiscal policy impact on poverty and inequality.

Stephanie Eckman is a fellow at RTI International in Washington, DC specializing in methods to collect high quality survey data. Her research focuses on the combination of survey and geospatial data. Previously, she held teaching and research positions at the Institute for Employment Research in Nuremberg, Germany and at the University of Mannheim. Dr. Eckman received a Ph.D. in survey methodology from the University of Maryland.

Alvin Etang is a senior economist in the Poverty and Equity Global Practice at the World Bank. Before joining the World Bank, he was a postdoctoral associate at Yale University. His interest in micro-development has led him to focus on analysis of poverty and welfare outcomes. With substantial experience in household survey design and implementation, Alvin has worked in several African countries. He is currently managing the World Bank's "Listening to Africa" initiative, mobile phone panel surveys for welfare monitoring, which has won many awards including for innovation and knowledge. He has also taught undergraduate economics courses, and has designed and used economic experiments as a tool to analyze poverty issues. His research has been published in several academic journals and has also featured in popular press such as *The Economist, Wall Street Journal, Financial Times, The Atlantic, Frontline*, among others. He is a co-author of the book titled *Mobile Phone Panel Surveys in Developing Countries: A Practical Guide for Microdata Collection*. He holds a Ph.D. in economics from the University of Otago in New Zealand.

Saad Gulzar is an Assistant Professor of Political Science at Stanford University. He uses field experiments and data from government programs to study the determinants of political and bureaucratic effort

Notes on Contributors xvii

toward citizen welfare. His research interests lie in the political economy of development and comparative politics, with a regional focus on South Asia. Gulzar earned his Ph.D. from New York University in 2017.

Kristen Himelein is a senior economist/statistician in the Poverty and Equity Global Practice at the World Bank, with extensive experience working in fragile, climate-affected, and post-conflict states. Her areas of expertise are survey methodology, sampling, and statistics, and her work has been published in peer-reviewed journals including the *Journal of Development Economics, Journal of Official Statistics*, and *Statistical Journal of the International Association for Official Statistics*, among others. She was also the project lead for high frequency cell phone surveys to measure the socio-economic impacts of Ebola in Sierra Leone and Liberia, which were widely disseminated in the international press. She holds a Master of Public Administration in International Development degree from the Harvard Kennedy School, and a graduate certificate in survey sampling from the Joint Program on Survey Methodology at the University of Maryland.

Johannes Hoogeveen is a lead economist in the Poverty and Equity Global Practice at the World Bank. He combines analytical and strategic work with the implementation of lending operations. He published academic papers on poverty measurement, survey design, statistics governance, education, nutrition, informal insurance, and land reform. His current research interest evolves around creating feedback loops (particularly in fragile situations exploiting new and established data collection technologies) and the relation between poverty, governance and identity. He was a manager at Twaweza, a national NGO in Tanzania, where he led a unit strengthening citizen accountability through feedback mechanisms. He holds a Ph.D. in economics from the Free University in Amsterdam.

Mohammad Isaqzadeh is a Ph.D. candidate at Princeton University. He has over seven years of experience as a consultant for the World Bank, working on the impact evaluation of NSP, NERAP, UCT and TUP programs in Afghanistan. He also taught for five years at the American University of Afghanistan. His research focuses on

xviii Notes on Contributors

insurgencies, post-conflict governance, and the role of religion in political mobilization and public goods provision. He has co-authored *Policing Afghanistan: The Politics of the Lame Leviathan* (Oxford University Press), and "Violence and Risk Preference: Experimental Evidence from Afghanistan" (American Economic Review). He holds a master's degree in international development from Oxford University.

Lennart Kaplan is a researcher at Göttingen and Heidelberg University. As a member of the research group "Globalization and Development" Lennart focuses on the meso-level of development research. More specifically, his research combines impact evaluation methods with geospatial and survey approaches.

Roy Katayama is a senior economist in the Poverty and Equity Global Practice at the World Bank. His current work focuses on the design of data collection methods suitable for fragile settings, performance-based financing for statistical capacity building, iterative beneficiary monitoring for improved project implementation, enhanced digital census cartography, geospatial analysis of development, and global poverty monitoring. During his time at the World Bank, he has led analytical work on poverty and inequality, poverty measurement, poverty maps, welfare impact of shocks, targeting of social safety nets, and systematic country diagnostics. He has experience working in numerous Sub-Saharan African countries. He holds a Master of Public Administration in International Development from the Harvard Kennedy School of Government.

Nandini Krishnan is a senior economist in the Poverty and Equity Global Practice of the World Bank, currently leading its Afghanistan program. In the past, she has worked as the poverty economist in Iraq and the Philippines, co-led a multi-country survey and analysis of host communities and Syrian refugees, and has supported regional and corporate initiatives for data and monitoring. She has worked on labor market, gender and inclusion issues in Egypt, Jordan, the Palestinian territories, Yemen, and the MENA region, and supported impact evaluations of large scale projects and programs in Tanzania, Nigeria, and South Africa. As a member of the World Bank Research Group's Social

Observatory Initiative, she supports World Bank operations to design systems that can learn from implementation data to improve effectiveness, and adapt program design. She holds a Ph.D. in Economics from Boston University.

Johan Mistiaen a Belgian national, joined the World Bank in 1999 and is currently the Program Leader and Lead Economist for Eritrea, Kenya, Rwanda and Uganda. He is based in Nairobi where he coordinates and supports the Bank's team responsible for delivering the analytical and operational portfolio managed by the Equitable Growth, Finance and Institutions group of Global Practices. He previously led the Bank's socio-economic and demographic data team and worked in the Bank's Research Department for some years. Johan studied Biology, Economics and Statistics at the Universities of York (UK) and Maryland (USA).

Juan Muñoz is the founder and managing partner of Sistemas Integrales, a firm created in 1970 and based in Santiago, Chile. He is interested in the application of statistics and computer science to economics, health, education and agriculture. As a consultant for universities, governments, international agencies and private clients, he has assisted in the design, implementation, steering and analysis of censuses and agriculture, budget, consumption, demographic, living standard, labor, and opinion surveys in over a hundred countries. These projects usually entail sampling and questionnaire design, survey organization and logistics, integration of computers to fieldwork, quality monitoring, report generation, and database documentation and dissemination.

Utz Pape is a senior economist in the Poverty and Equity Global Practice at the World Bank. He leads teams to design and implement lending projects to improve national statistical systems and to prepare analytical poverty work including poverty assessments, poverty impact studies, and Systematic Country Diagnostics. His work experience in post-conflict countries contributes to his research agenda including the design of methodologies for poverty measurement in fragile settings. His research has received awards and is published in peer-reviewed journals, including *Nature*. He holds a Ph.D. from the International Max

Planck Research School and the Free University of Berlin and was a postdoctoral associate at Harvard University. He also holds a Master of Public Administration/International Development from the London School of Economics and the School for International and Public Affairs (SIPA) at Columbia University.

Flavio Russo Riva is a Ph.D. candidate in Government and Public Administration at the São Paulo School of Administration. His research focuses on impact evaluation of public policies and social programs in Brazil's public education and health systems using observational data and randomized controlled trials. He has worked as a short-term consultant for the Poverty and Equity Global Practice at the World Bank and the Inter-American Development Bank in the last years.

Jacob Shapiro is a Professor of Politics and International Affairs at Princeton University and directs the Empirical Studies of Conflict Project, a multi-university consortium that compiles and analyzes micro-level data and other information on politically motivated violence in countries around the world. He is author of *The Terrorist's Dilemma: Managing Violent Covert Organizations* and co-author of *Small Wars, Big Data: The Information Revolution in Modern Conflict*. His research has been published in broad range of journals in economics and political science as well as a number of edited volumes. He has conducted field research and large-scale policy evaluations in Afghanistan, Colombia, India, and Pakistan. Shapiro received the 2016 Karl Deutsch Award from the International Studies Association, given to a scholar younger than 40 or within 10 years of earning a Ph.D. who has made the most significant contribution to the study of international relations.

Dhiraj Sharma is an economist in the Poverty and Equity Global Practice. His work focuses on welfare measurement, poverty diagnostics, and policy analysis. He has led or contributed to the analysis of poverty in Ghana, Iraq, and Nepal, and has led impact evaluations in Nepal. His current work focuses on welfare analysis and statistical capacity building in countries in the Middle East and North Africa region. His recent work in the region includes research on the impact of refugee influx on host communities. He is a co-author of the *Poverty and Shared*

Prosperity 2018: Piecing Together the Poverty Puzzle, the World Bank's biennial publication on global extreme poverty. Dhiraj holds a Ph.D. in applied economics from the Ohio State University.

Andre-Marie Taptué is an economist in the Poverty and Equity Global Practice at the World Bank. He developed and implemented the Beneficiary Monitoring (IBM) System. He has also led the Permanent Monitoring System in North Mali and is supporting implementation of the Third-Party Monitoring in Mali. He is currently working on analytical work, a statistical project, policy dialogue, and extending IBM. Prior to joining the Bank, Andre-Marie was a lecturer at Laval University in Canada. He also worked as an economist statistician at the Department of Studies and Statistical Surveys of the National Institute of Statistics in Cameroon. He earned a Ph.D. in economics at Laval University and a master's degree in statistics and economics at ISSEA in Cameroon.

Tara Vishwanath is currently a lead economist in the Europe and Central Asia region's Poverty and Equity Practice of the World Bank and Global Lead on Welfare Implications of Climate, Fragility and Conflict Risks. She has led numerous analytical products on poverty, inequality and employment in countries in the South Asia and Middle East and North Africa; more recently co-leading the multi-topic survey and analysis of Syrian refugee and host communities in Lebanon, Jordan and Northern Iraq. Before joining the World Bank, she was a Professor in the Department of Economics at Northwestern University and has published widely in refereed economics journals spanning research topics in economic theory, labor economics and development. She holds a Ph.D. in Economics from Cornell University.

James Walsh is a member of the World Bank's Behavioral Science Unit, eMBeD, and a doctoral student at the Blavatnik School of Government at the University of Oxford. He was a member of the research team for the World Development Report 2015: Mind, Society, and Behavior and served on the faculty of the Georgetown School of Foreign Service where he lectured in behavioral approaches to development economics. He holds a B.A. in Economics and Political Science

xxii   Notes on Contributors

from Trinity College Dublin and a Master of Public Policy from the Kennedy School of Government at Harvard University.

Gervais Chamberlin Yama is a statistician in the Poverty and Equity Global Practice at the World Bank, with experience in working in fragile and conflict-afflicted states. He has extensive experience in designing executing and managing surveys in the Central African Republic, Democratic Republic of Congo, and the Republic of Congo. He has recently developed a new approach to performance-based data collection for enumerators and supervisors in the Central African Republic that enhances data quality and promotes efficiency. He holds a master's degree in statistics from the Sub-Regional Institute of Statistics and Applied Economics (ISSEA) in Yaoundé, Cameroon.

List of Figures

Chapter 1

Fig. 1 Extreme poverty (2017 or latest available number)
(*Source* World Bank, Poverty and Equity Data Portal,
accessed November 2017)   2

Chapter 2

Fig. 1 Timing of Sierra Leone and Liberia high-frequency mobile
phone surveys (Color figure online) (*Note* Shading reflects
dates of data collection. *Source* Authors calculations based
on WHO Sit Rep data)   16

Fig. 2 Responses on food security issues from various L2A surveys
(*Source* Authors' calculations from the Malawi, Madagascar,
and Senegal L2A surveys)   18

Fig. 3 Evidence from the Sierra Leone and Liberia phone surveys
(*Source* Sierra Leone high-frequency mobile phone survey
and Liberia high-frequency mobile phone survey)   21

Fig. 4 Response rates for the high-frequency mobile phone surveys
in Sierra Leone and Liberia (*Source* Authors' calculations)   24

xxiv   List of Figures

Chapter 3

Fig. 1 Food security score by country (*Source* Author's calculations based on RERS data)   38

Fig. 2 Trends in income and food storage (*Source* Author's calculations based on RERS Nigeria data [Conducted with the National Bureau of Statistics of the Federal Government of Nigeria under Poverty and Conflict Monitoring Systems])   41

Fig. 3 Trends in livestock losses and employment (Somalia) (*Source* Author's calculations based on Somali RERS)   42

Fig. 4 Challenges in accessing food markets and livestock losses (South Sudan) (*Source* Author's calculations based on RERS South Sudan)   42

Fig. 5 Challenges in accessing food markets and reasons for low school attendance (Yemen) (*Source* Author's calculations based on RERS Yemen)   43

Fig. 6 Survey duration and implementation costs (*Source* Author's calculations based on RERS)   44

Fig. 7 Proposed adaptive questionnaire design (Color figure online)   46

Fig. 8 Proposed timeline for adaptive questionnaire design with RERS methodology   47

Chapter 4

Fig. 1 Population pyramids before and after the 2012 crisis (*Source* Mali census data for 2009, INSTAT 2012; authors' calculations using January 2016 Permanent Monitoring baseline survey)   52

Fig. 2 Timing of return (percentage) (*Source* Authors' calculations based on the Mali Listening to Displaced People Survey)   54

Fig. 3 Level of education of population aged 18+ (percentages) (*Source* Authors' calculations using the Listening to Displaced People Survey and the Enquête Modulaire et Permanente, EMOP 2011, of the Mali National Institute for Statistics, INSTAT)   55

Fig. 4 Asset ownership compared with regional average (*Source* Authors' calculations using the Listening to Displaced People Survey, 2014 and the Enquête Modulaire et Permanente, EMOP, 2011 of the Mali Institute of Statistics (INSTAT))   57

List of Figures   **xxv**

Fig. 5 Changes in perceived living conditions over the duration
of the survey (*Source* Authors' calculations based on Mali
Listening to Displaced People Survey)   58

Fig. 6 Fixed effects regression on the decision to return
(*Source* Hoogeveen et al. 2019)   59

Fig. 7 Attrition rates (*Source* Authors' calculations using
the Mali Listening to Displaced People Survey)   60

Chapter 5

Fig. 1 Map indicating the location of enumerators across northern
Mali (*Source* World Bank 2016)   70

Fig. 2 Percentage of households living in a state of food insecurity
(*Source* Authors' calculations based on data from
the Permanent Monitoring System)   73

Fig. 3 Perceptions of security (*Source* Authors' calculations based
on Mali Permanent Monitoring System)   73

Fig. 4 Confidence in the government and the judicial system
(*Source* Authors' calculations based on Mali Permanent
Monitoring System)   74

Fig. 5 Confidence in people (*Source* Authors' calculations based
on Mali Permanent Monitoring System)   75

Fig. 6 Problems reported by health facilities and schools
(*Source* Authors' calculations based on Mali Permanent
Monitoring System)   76

Chapter 6

Fig. 1 Selected results from the district census (*Source* Authors'
calculations based on the CAR District Census/ENMC)   90

Fig. 2 Selected results on local administration, infrastructure,
and access to services (*Source* Authors' calculations based
on the CAR District Census/ENMC)   91

Fig. 3 Local Development Index across districts (*Source* Authors'
calculations based on the CAR District Census/ENMC)   93

Fig. 4 Food consumption by wealth and agro-ecological zone
(*Source* Authors' calculations based on the CAR District
Census/ENMC)   94

xxvi   List of Figures

Fig. 5   Mali Local Development Indices, by region and livelihood
         zone (*Source* Authors' calculations based on the Mali
         Commune Censuses)   98
Fig. 6   LDI 2006 and 2017 by region and 2017 LDI map (*Source*
         Authors' calculations based on the Mali Commune Censuses)   98

Chapter 7

Fig. 1   Building classifications (Color figure online) (*Source* Authors'
         calculation)   107
Fig. 2   Boundaries and population densities (*Source* Authors'
         calculation)   109
Fig. 3   Listing totals, modeled estimates, and rooftop counts
         for Makala (*Source* Authors' calculations)   111
Fig. 4   Stratification map   114
Fig. 5   Viewshed analysis (Color figure online)   118
Fig. 6   Size and location of selected PSUs   120
Fig. 7   Example of the Qibla method   121
Fig. 8   Mean and confidence intervals (by method)   122

Chapter 9

Fig. 1   Illustration of the rapid consumption survey methodology
         (using illustrative data only)   156
Fig. 2   Average relative bias and standard error   160
Fig. 3   Bias and standard errors   162
Fig. 4   Cumulative consumption distribution (in USD) per day
         and per capita (Color figure online) (*Note* For core module
         (dark blue), core and assigned optional modules (medium
         blue), and imputed consumption (light blue). The presented
         consumption aggregate does not include consumption
         from durable goods   166

Chapter 11

Fig. 1   Treatment Components (*Source* Authors' visualization)   196
Fig. 2   Consumption distribution by population and treatment
         (*Source* Authors' calculations using HFS 2017, IDPCSS
         2017 and CRS 2017)   199

Fig. 3	Number of items consumed by population and treatment (*Source* Authors' calculations using HFS 2017, IDPCSS 2017 and CRS 2017)	199
Fig. 4	Treatment effects across quintiles (IDPs) (*Source* Authors' calculations using HFS 2017, IDPCSS 2017 and CRS 2017. All regressions use clustered robust standard errors [White 1980]. Confidence bands refer to the 95% confidence interval. Consumption quantities, values, and calories are used in per-adult equivalent terms. The regression framework is introduced in the appendix. No sampling weights are used as 'honesty primes' are expected to affect, specifically, the extremes of the distribution and the average treatment effect is not a priori of interest)	201
Fig. 5	Treatment effects across quintiles (non-IDPs) (*Source* Authors' calculations using HFS 2017, IDPCSS 2017 and CRS 2017. All regressions use clustered robust standard errors [White 1980]. Confidence bands refer to the 95% confidence interval. Consumption quantities, values, and calories are used in per-adult equivalent terms. The regression framework is introduced in the appendix. No sampling weights are used as 'honesty primes' are expected to affect, specifically, the extremes of the distribution and the average treatment effect is not a priori of interest)	202

Chapter 13

Fig. 1	Five steps of the IBM approach	220
Fig. 2	Small samples may suffice to uncover problems (*Source* Uwazi 2010)	222
Fig. 3	Regular follow-up improved school feeding performance (*Source* Authors' calculations based on IBM data)	225
Fig. 4	Selected gender outcomes uncovered by different IBM activities (*Source* Hoogeveen et al. 2018)	230

List of Tables

Chapter 1

Table 1   Topical guide to this book   9

Chapter 3

Table 1   Developmental gaps among the famine-risk population   39

Chapter 5

Table 1   Response rate (percentage of households that answered the survey)   71

Chapter 6

Table 1   Local Development Index: components and weights   89

Chapter 7

Table 1   Stratification of the Afar region   115

Chapter 8

Table 1	Syrian Refugee and Host Community Survey: sampling strata—Lebanon	143
Table 2	List of selected segments (enumeration areas)—Lebanon	144
Table 3	List of sample super segments (for CFs divided into super-segments or secondary sampling units)—Lebanon	147

Chapter 9

| Table 1 | Number of items and consumption share captured per module | 159 |
| Table 2 | Number of items and consumption shares captured per module | 165 |

Chapter 10

| Table 1 | Survey approaches and addressing sensitivity biases | 177 |

Chapter 11

| Table 1 | Results using poverty thresholds | 202 |

Chapter 13

| Table 1 | Iterative feedback approach for a school feeding project in Mali | 227 |
| Table 2 | Iterative feedback for a project distributing agricultural inputs using electronic vouchers | 229 |

Chapter 14

| Table 1 | Resource requirements to implementing methods described in various chapters | 238 |

List of Boxes

Chapter 1

Box 1   Using tablets for data collection allows for a rich array of innovations   10

Chapter 6

Box 1   LDI in Mali allows for comparisons across time and space   96

Chapter 13

Box 1   Beneficiary monitoring is not a new concept, but light monitoring is   218

Box 2   How IBM compares to project monitoring   222

1

Fragility and Innovations in Data Collection

Johannes Hoogeveen and Utz Pape

1 Introduction

Fragility, conflict, and violence (FCV) represent a critical development challenge that threatens efforts to end extreme poverty and promote shared prosperity. Two billion people live in countries where development outcomes are affected by FCV, including many countries in Africa. Of the 38 countries on the World Bank's official 2018 FCV list, 20 can be found in Africa. Moreover, while the global share of the extreme poor living in conflict-affected situations is about 20%, this number is much higher in Africa, around 32%. In fact, nearly 80% of all poor people living in conflict-affected situations reside in Africa (Fig. 1).

J. Hoogeveen (✉) · U. Pape
World Bank, Washington, DC, USA
e-mail: jhoogeveen@worldbank.org

U. Pape
e-mail: upape@worldbank.org

© International Bank for Reconstruction and Development/The World Bank 2020
J. Hoogeveen and U. Pape (eds.), *Data Collection in Fragile States*,
https://doi.org/10.1007/978-3-030-25120-8_1

1

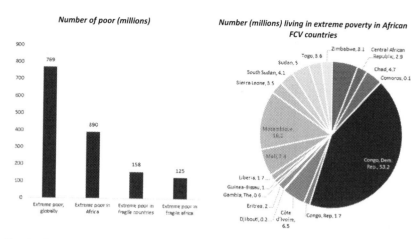

Fig. 1 Extreme poverty (2017 or latest available number) (*Source* World Bank, Poverty and Equity Data Portal, accessed November 2017)

Particularly worrisome is that between now and 2030, the share of extremely poor people living in FCV countries is expected to rise from 20 to 50%. Given that most of these people are likely to be in Africa, it is unsurprising that at the 2015 Annual Bank Conference on Africa, Makhtar Diop, the then World Bank's vice president for the region, emphasized not only the importance of fragility, but also the need for a much more profound inquiry into its drivers and consequences: "Conflict and fragility exact a costly toll on the economies of Africa. As we scale up our operational work in fragile states, a better understanding of the causes and impacts of conflict and fragility can help to prevent some of the deadly conflicts at the community level."

A better understanding of socio-economic well-being of citizens in such countries as well as measuring the impacts of shocks and conflicts start with better data. Data deprivation is a pressing problem in FCV settings for both decision makers and its citizens, and in particular, for the poor, who often lack voice and agency, and who may remain invisible unless data identify their existence and state of being. The need for reliable data on living conditions in fragile situations is even greater, and yet data deprivation tends to be worse in such contexts. Data can provide evidence on the plight of some of the most vulnerable populations,

1 Fragility and Innovations in Data Collection

such as the displaced, or those affected by natural disasters, violence, famine, or epidemics, and can facilitate the formulation of policy responses by decision makers. As such, there is an urgent need for data in fragile situations.

This book attempts to address this data challenge. It reflects work carried out by World Bank staff from the Poverty and Equity Global Practice and by others covering our experiences in fragile situations, facing challenges around data collection, mostly in Africa in the Central African Republic, the Democratic Republic of Congo, Liberia, Madagascar, Mali, Malawi, Nigeria, Senegal, Sierra Leone, Somalia, South Sudan but also in Iraq, Jordan, Lebanon, and Yemen.[1] Typical welfare surveys such as the Living Standard Measurement Surveys (LSMSs) and Household Budget Surveys (HBSs) that are implemented in a large number of countries are not always appropriate for these situations. Because of the pressing demand for data, there has been significant support for experimentation and innovation around data collection methods. This has allowed us to develop solutions suitable for these contexts, which are often equally relevant for non-fragile settings.

Through our experiences in identifying innovative ways to collect data, we have learned three lessons. First, it is possible to collect high-quality data in fragile settings. Doing so may require adaptations to the data collection process but situations in which no information can be collected are rare. Second, data collection in fragile contexts does not need to be more expensive than in other settings. In fact, the costs associated with many of the innovations discussed in this book compare favorably to more traditional data collection methods. Third, a careful assessment of the data needs of decision makers is essential. Often relatively easy-to-collect information goes a long way toward meeting their demands, as long as it is provided in a timely fashion. This holds

---

[1]Not all these countries are on the fragile country list maintained by the World Bank and downloadable from: http://www.worldbank.org/en/topic/fragilityconflictviolence/brief/harmonized-list-of-fragile-situations. When countries are not on the fragile country list, the discussed approaches were typically applied during an emergency, as was the case during the Ebola crisis in Sierra Leone and the 2016 floods in Malawi.

particularly in volatile situations. Hence it may be sufficient to demonstrate whether respondents can engage in certain income-generating activities, without measuring how much income is actually earned. Perception questions, eliciting information about trust, security, or development priorities, tend to be very informative for decision makers in unstable settings where rumors spread quickly and where opinion polls and (objective) media reporting are absent. In other instances, simple to collect information does not suffice. We present such a case in Chapter 9 for Somalia where estimates of poverty had to be produced even though interviews could not be lengthy for security reasons, precluding asking detailed consumption questions.

There was also a fourth lesson: technology is not a panacea for all data collection issues and not everything works. We considered machine learning and big data, but these approaches were not successful. Cloud computing and improvement of statistical learning algorithms enable the use of satellite images and other sources of big data, but satellite images can be expensive, the methodologies can be complex, and external validity is at times difficult to ensure. Some data collection exercises were discontinued because of a lack of funding (and by implication, a lack of demand). Tablets facilitated electronic data collection and reduced field supervision, but in some situations, its use complicated data collection as it raised suspicion from respondents or unwanted attention from thieves. Improved mobile phone coverage also created the opportunity to use mobile phone interviews for data collection in insecure areas, but the resulting information may not be representative of the population.

It has been immensely rewarding to find ways to produce reliable data in the face of significant challenges: absent sampling frames, high levels of insecurity, and limited budgets. We feel privileged to have been given the opportunity to collect data that has helped inform decision makers at critical junctures of the development process. However, we also realize that our work is far from complete. With adaptations, many of the innovations presented in this book are scalable. This holds for district censuses, which are highly suited to inform decentralization processes, or Iterative Beneficiary Monitoring (IBM), which can be used to improve project performance in any context. Rapid consumption surveys have the potential to significantly reduce the cost of collecting consumption data, and sampling frames derived from satellite images can be used more systematically to update sampling frames. Moreover, with cell phone coverage continuously improving, mobile phone surveys (examples presented in this book are monitoring the Ebola crisis and people displaced by the crisis in Mali, and to inform a famine response in Nigeria, Somalia, South Sudan, and Yemen) that can be scaled up rapidly during a crisis deserve to become part of the regular tool-box of disaster planning, as they can offer timely data when a crisis is imminent.

2 Data Collection in Fragile Situations

Fragility, conflicts and violence affect data collection in multiple ways. The capacity to implement and analyze complex surveys tends to be limited and resources to pay for data collection are scarce as the revenue generating capacity in FCV settings tends to be constrained and because funding for data collection competes with other urgent needs. For these reasons, few household surveys are implemented in fragile situations, or if they are, are not implemented regularly or without covering the entire territory. In addition, risks in FCV countries are oftentimes elevated, because of violence but also because of other dangers, such as disease. In Somalia, for instance, a traditional household consumption survey with interview lengths exceeding several hours was not possible given the level of insecurity and danger imposed to enumerators if spending more than one hour with a household. During the Ebola crisis, enumerators could not travel and collect information from respondents using face-to-face interviews because of the risk of infection.

Data collection during conflict is also affected by poor road quality, inadequate telecommunications infrastructure and, at times, populations that are hostile to representatives of the central government offering little in terms of key public services. The reason for these challenges is because conflicts tend to occur in locations that are physically distant from administrative centers, isolated, have low population density and few key public services, and which bear the brunt of weak state capacity.

Collecting data in such situations is not only logistically challenging, but people living in these areas often feel little loyalty to the distant capitals that have historically ignored them and may be hostile to anyone seen to represent the state.

Mobile target populations are a further complication often associated with data collection in fragile situations. Mobility is a challenge not only because pastoralists tend to live in distant, low-density areas that are often the theaters of conflict, but also because displacement is a major issue during times of insecurity. During the crisis in northern Mali, for example, 36% of the population fled the area, and in the Central African Republic, 25% of the population was displaced. The United Nationals High Commissioner for Refugees (UNHCR) estimated that by the end of 2016, there were 5.1 million refugees in Africa, with the Central African Republic, the DRC, Somalia, South Sudan, and Sudan being the major sources of refugees. The number of internally displaced people (IDPs) is even higher, with almost 9 million displaced people between these five countries alone.

Data collection in FCV settings is also affected by the absence of adequate sampling frames, which may have been lost or are simply out of date. In the case of the Central African Republic, for instance, during the civil war, much of the data infrastructure (buildings, books, maps, servers, and computers) was lost to looting. However, even without the looting, sampling frames would no longer have been valid as a large proportion of the population had become displaced. Finally, there is often time pressure, as decision makers require accurate information with a quick turnaround. In the Central African Republic, following the signing of the Peace Accord, the team had 90 days to prepare, field, and analyze a survey to yield representative data on the development priorities of citizens. The pressure to inform decision makers during or directly after a disaster can be even higher, for example, in the Ebola-affected countries, or for the drought response in Nigeria, Somalia, South Sudan, and Yemen.

Because traditional data collection methods are not always suited to fragile situations, this book presents innovations developed to deal with some of these challenges. Some, though not all, were also motivated by the fact that data needs in fragile situations are different. There is much more emphasis on timely data that can monitor a given situation than on in-depth analyses to inform policy decisions. For example, policy-makers in insecure settings often prefer knowing where schools are and whether they are still functioning, rather than seeing a detailed analysis of whether the rate of return to education is higher at the primary or tertiary level. This reality has shaped some of the data collection processes presented in this book, as questionnaires these contexts can be less comprehensive. This in turn can be effectively combined with mobile phone interviews as a data collection method, which typically should not last longer than 20–30 minutes, and interviews by locally resident enumerators who cannot be retrained for every new questionnaire. District surveys introduced in the Central African Republic and Mali capitalized on the realization that an index reflecting the degree of public service provision (health, water, education, and infrastructure) at the lowest administrative level was a pragmatic alternative to a more detailed poverty map, which would take a long time to create. The IBM approach introduced in Mali which offers feedback to project staff drawn from light data collection exercises, was developed to complement project supervision missions, which had become difficult to conduct due to security concerns. The approach relies on highly simplified data collection tools, which ensure focus, speed and allow to keep cost down.

Simplifications, are not always possible. In Somalia, for instance, up-to-date poverty estimates were needed to inform the Heavily Indebted Poor Countries (HIPC) process. Under normal circumstances, estimating poverty requires administering a lengthy consumption module that takes several hours to complete. However, due to security concerns, it was advised that the maximum duration of a household interview should not exceed 60 minutes. This time restriction meant that a lengthy consumption module was not possible, even if questions about education, health, and perceptions were dropped. Using a new questionnaire design with smart sampling techniques at the level of questions solved this challenge.

To structure the book, we organized it into three parts. Part I: "Innovations in Data Collection" presents ways to collect data that are cognizant of security and other risks, as well as the specific data needs of decision makers in FCV countries. The first three chapters in this section discuss data collection using mobile phone interviews. Chapter 2 provides an example of this method during the Ebola crisis in Sierra Leone. Chapter 3 describes how mobile phone interviews were used to inform a response to the drought in Nigeria, Somalia, South Sudan, and Yemen. Chapter 4 reports an exercise to track people displaced by the crisis in northern Mali. Chapter 5 discusses how, in situations where travel by outsiders is too dangerous, data collection may still be feasible by relying on locally recruited, resident enumerators who are trusted by their community. Chapter 6 discusses the district survey and Local Development Index introduced in the Central African Republic. It informed the Recovery and Peace Building Assessment and collects much of the data that feeds into the national monitoring system.

Part II: "Methodological Innovations" presents innovations with respect to collecting data and sampling. To deal with the absence of sampling frames in the DRC and Somalia, satellite images and sophisticated machine learning algorithms were used to estimate population density and demarcate enumeration areas (Chapter 7). The same chapter also showcases a novel sampling approach implemented in the Afar region of Ethiopia to ensure that pastoralists were adequately included. This approach was also used in Somalia to avoid listing exercises that were viewed with suspicion by community and authorities. Chapter 8 discusses sampling for representative surveys of displaced populations, using the example of Syrian refugees and host communities in Jordan, Lebanon and Kurdistan, Iraq. Chapter 9 offers a solution for those interested in collecting poverty estimates for insecure locations in which the time available for face-to-face interviews is too limited to implement lengthy household consumption expenditure surveys that are generally used for measuring poverty. Chapters 10 and 11 discuss how to elicit truthful information from respondents. Chapter 10 focuses on asking questions about sensitive issues such as e.g. loyalty to controversial groups while Chapter 11 deals with how to avoid strategic responses when respondents might expect benefits to be associated with certain answers.

Part III: "Other Innovations" presents a project that used video testimonials (Chapter 12) as a unique and cost-effective way to give external

Table 1 Topical guide to this book

Chapter topic	Collecting data unsafe environments	Monitoring of crises, projects and country programs	Asking sensitive questions	Dealing with inadequate sampling frames	Following mobile populations	Collecting data rapidly	Promoting development outcomes	Repeated data collection and panel data
Mobile phone surveys	✓	✓				✓		✓
Rapid reponse survey	✓	✓				✓		
Tracking displaced people					✓			✓
Resident enumerators	✓					✓		✓
Local development index		✓				✓		✓
Geo spatial sampling				✓				
Sampling displaced populations				✓				
Rapid consumption surveys	✓					✓		
Studying sensitive topics			✓					
Eliciting accurate responses			✓					
Video testimonials							✓	
Iterative beneficiary monitoring		✓				✓		

audiences a perspective on the lives of survey respondents. In South Sudan, a web portal was created where one can watch short video testimonials of respondents describing their situation in their own words, which not only provided the necessary context for the quantitative results, but also gave a voice to the poor. In Chapter 13, IBM is discussed, which relies on light-touch, repeated data collection exercises to create dynamic feedback loops for project staff. IBM has been found to enhance the efficiency of projects and is, because of its minimalist data demands, highly suited for fragile contexts.

We have aimed to keep this book practical and accessible, focusing on illustrations and applications, as our objective is to provide the reader with examples of what is feasible. Every chapter presents the data challenge, how it was addressed, and lessons learned. For readers interested in specific topics, we present in Table 1 an overview of which chapters might be of interest. For example, if the concern is that respondents might give biased answers, because questions touch upon sensitive issues or because the respondent may believe that the right responses can result in certain benefits, then Chapters 10 and 11, which discuss methodological solutions and behavioral nudges respectively would be worth reading.

Box 1   Using tablets for data collection allows for a rich array of innovations

Using tablets or mobile phones to collect data, or more specifically, Computer-Assisted Personal Interviews (CAPI), led to more changes than making data entry obsolete. Enumerator error can be reduced with dynamic validity checks and complex skipping patterns, opening up new possibilities. The randomization of questions can now be automated, for instance, a feature that has been part of rapid consumption surveys (Chapter 9) and list experiments (Chapter 10). Complex survey skipping patterns, not possible in paper questionnaires, become an additional option.

To improve accuracy, CAPI can identify implausible responses and request enumerators to verify or correct their responses before proceeding. This has proved useful in consumption modules, where responses can be assessed against caloric needs, or where unit values can be checked

1 Fragility and Innovations in Data Collection 11

against plausible price ranges. Photos can also be used to obtain more reliable estimates of otherwise hard to quantify, and seasonably variable units such as a "heap" or "bunch."

The use of tablets also improves supervision. GPS locations can be collected in the background, allowing supervisors to assure that enumerators are where they are expected to be, and also assess the spatial distribution of a sample. Tablets can monitor the time it takes to record answers, and interview snippets can be recorded randomly. These features can quickly confirm whether interviews are actually conducted, reducing the need for unannounced supervision visits.

Enumerators can also take advantage of the additional hardware included in tablets. For panel surveys, households can be given a barcode, which can be photographed or scanned with a tablet, thus reducing the frequency of mistakes. The ability to take pictures and shoot video can be used to enrich feedback in other ways as well. Chapter 12 presents an instance where enumerators were trained to use their tablets to record—after the formal interview—stories about the experiences of interviewees.

Where mobile phone networks are available, tablets can send data for aggregation and real-time analysis, significantly reducing the time it takes to produce results. As data is typically sent into the "cloud," such analysis can be done anywhere across the globe. The Rapid Emergency Response Survey presented in Chapter 3 made use of this feature. When enumerators are in the field for a long time, or when questionnaires needed to be updated because errors need to be corrected, the use of tablets allows for remote questionnaire management, a feature used in Chapter 5 to provide resident enumerators with new survey instruments and questions.

The opinions expressed in this chapter are those of the author(s) and do not necessarily reflect the views of the International Bank for Reconstruction and Development/The World Bank, its Board of Directors, or the countries they represent.

Open Access This chapter is licensed under the terms of the Creative Commons Attribution 3.0 IGO license (https://creativecommons.org/licenses/by/3.0/igo/), which permits use, sharing, adaptation, distribution and reproduction in any medium or format, as long as you give appropriate credit to the International Bank for Reconstruction and Development/The World Bank, provide a link to the Creative Commons license and indicate if changes were made.

Any dispute related to the use of the works of the International Bank for Reconstruction and Development/The World Bank that cannot be settled amicably shall be submitted to arbitration pursuant to the UNCITRAL rules. The use of the International Bank for Reconstruction and Development/The World Bank's name for any purpose other than for attribution, and the use of the International Bank for Reconstruction and Development/The World Bank's logo, shall be subject to a separate written license agreement between the International Bank for Reconstruction and Development/The World Bank and the user and is not authorized as part of this CC-IGO license. Note that the link provided above includes additional terms and conditions of the license.

The images or other third party material in this chapter are included in the chapter's Creative Commons license, unless indicated otherwise in a credit line to the material. If material is not included in the chapter's Creative Commons license and your intended use is not permitted by statutory regulation or exceeds the permitted use, you will need to obtain permission directly from the copyright holder.

Part I
Innovations in Data Collection

2

Monitoring the Ebola Crisis Using Mobile Phone Surveys

Alvin Etang and Kristen Himelein

1 The Data Demand Challenge

The outbreak of the Ebola virus disease in West Africa in 2014 constituted one of the gravest global health emergencies of recent years.[1] The Ebola outbreak originated in rural Guinea in December 2013, and then spread across the country and to the neighboring countries of Liberia and Sierra Leone. The pandemic continued for two years and the World Health Organization (WHO) only declared Liberia free of Ebola in May 2015, Sierra Leone in November 2015, and Guinea in December 2015. By the end of the crisis, the epidemic had claimed more than

---

[1]Henceforth, the term Ebola is used to refer to the virus, the disease, or the epidemic outbreak.

A. Etang (✉) · K. Himelein
World Bank, Washington, DC, USA
e-mail: aetangndip@worldbank.org

K. Himelein
e-mail: khimelein@worldbank.org

© International Bank for Reconstruction and Development/The World Bank 2020
J. Hoogeveen and U. Pape (eds.), *Data Collection in Fragile States*,
https://doi.org/10.1007/978-3-030-25120-8_2

15

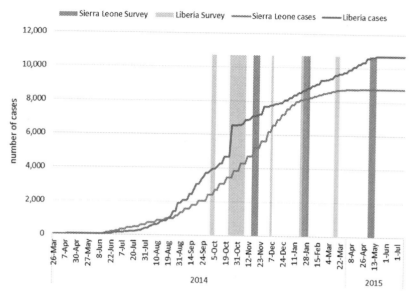

Fig. 1 Timing of Sierra Leone and Liberia high-frequency mobile phone surveys (Color figure online) (*Note* Shading reflects dates of data collection. *Source* Authors calculations based on WHO Sit Rep data)

11,300 lives in these three countries, including over 500 frontline healthcare workers.[2]

In addition to its effects on people's health, Ebola caused widespread economic disruption. At the height of the epidemic, schools, and markets were closed, government workers were placed on furlough, social gatherings were banned, transportation restrictions were placed on people and goods, and international borders were closed. Therefore, in addition to the health monitoring by the WHO, there was an urgent need for just-in-time data in order to monitor the economic impact of Ebola on livelihoods and wellbeing. Given the epidemic, however, it was impossible to deploy enumerators to the field to collect information from households and communities through face-to-face interviews.

---

[2]World Bank (2016).

The solution to this challenge came from the realization that the rapid spread of mobile phone coverage had created possibilities to monitor the crisis through mobile phone interviews. Mobile phones are particularly useful in situations in which data must be collected rapidly, at low cost, and/or in situations where traditional face-to-face interviews are not possible. In Sierra Leone and neighboring Liberia, it allowed for a timely response by providing critical data to decision makers about household welfare at the height of the crisis and during its aftermath (Fig. 1).

2 The Innovation

The proliferation of mobile phone networks and inexpensive handsets has opened up new possibilities for data collection. Since 2012, the Africa region of the World Bank supports a mobile phone survey initiative called Listening to Africa (L2A). L2A collaborates with statistical agencies and offers the possibility to complement face-to-face household surveys with mobile data collection.[3]

The standard L2A approach starts with a face-to-face household survey that serves as a baseline. This baseline survey ensures that the randomly drawn sample is representative of the target population. During this survey, each respondent receives a simple mobile phone and when necessary, a solar charger. The respondents then receive calls from a call center every month, which conducts the mobile phone interviews. Survey questions are programmed in computer-assisted telephone interview software, allowing questions to be posed, and answers to be simultaneously recorded. The phone interviews are short so that data can be collected quickly, and respondents do not become overly fatigued. Data, once collected, are made available to the public.

The L2A approach has been introduced in several countries, including Madagascar, Malawi, Mali, Senegal, Tanzania, and Togo, and the

---

[3]More information on this approach, including the instruments used, can be found on the L2A website: http://www.worldbank.org/en/programs/listening-to-africa. See also: Johannes Hoogeveen et al. (2014).

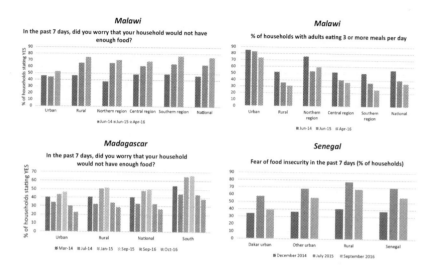

Fig. 2 Responses on food security issues from various L2A surveys (*Source* Authors' calculations from the Malawi, Madagascar, and Senegal L2A surveys)

L2A team has prepared a handbook documenting its experiences.[4,5] A two-minute video explaining the L2A approach can be found on the World Bank's website.[6]

While typical L2A questionnaires are fixed ahead of time, the instrument is flexible and can adapt to unforeseen needs. In particular, the high-frequency collection was well-suited to monitor food security, and the L2A team was able to respond to the unfolding situations in Malawi, Senegal, and Madagascar (Fig. 2). A sample questionnaire with food security questions that can be used for mobile phone interviews is presented in the annex to this chapter.

When the Ebola crisis began in 2014, the World Bank team had accumulated several years of experience with mobile phone surveys. Building on the L2A model, high-frequency mobile phone interventions were designed to provide rapid monitoring of the socio-economic impacts of

---

[4]Dabalen et al. (2016).

[5]Available at https://openknowledge.worldbank.org/bitstream/handle/10986/24595/9781464809040.pdf.

[6]http://www.worldbank.org/en/news/video/2017/01/23/listening-to-africa-a-new-way-to-gather-data-using-mobile-phones.

Ebola in Liberia and Sierra Leone.[7] As the L2A approach had shown, baseline information was needed to anchor estimates in a representative dataset. Fortunately, there were recent surveys in both countries that could serve this purpose. In Liberia, the Household Income and Expenditure Survey (HIES) was being conducted as the crisis broke out, and was forced to curtail its fieldwork in August 2014. Though only about half of the sample (4075 households) were surveyed, it was nationally representative, and despite not being planned as a panel survey, had collected phone numbers and contact information for respondents. Overall, 57% of HIES households reported a mobile phone number for at least one household member. This database of phone numbers and household characteristics became the sample frame for the mobile phone survey sample. In total, five rounds of phone interviews were completed between October 2014 and March 2015. Data were collected by the Gallup Organization from their US-based call centers, as there was no suitably experienced call center on the ground in Liberia, and it was not possible to bring in international experts due to the travel ban. While using an external call center posed several challenges, including a lack of proficiency in local languages, unwillingness of respondents to speak to strangers, and a high costs of calling, the survey was able to conduct 2781 interviews with 1082 unique households over the five rounds.

In Sierra Leone, the 2014 Labour Force Survey (LFS) was also being carried out during the Ebola crisis, with fieldwork completed in July 2014. The LFS is a nationally representative survey, with a sample size of 4188 households. It was planned as a panel survey, and had therefore collected phone numbers and contact information, with 66% of LFS households reporting a mobile phone number for a least one household member. Using this database, three rounds of data collection were completed between November 2014 and May 2015. Data were collected through a call center at the national statistics bureau, Statistics Sierra Leone, supervised by Innovations for Poverty Action for the first two rounds and supervised directly by the World Bank for round three. The survey was able to reach 2111 respondents over the three rounds (Himelein et al. 2015).

---

[7]A mobile phone survey was also conducted in Guinea but using a different methodology (World Bank (2016).

3 Results from the Ebola Surveys

The Ebola surveys covered a wide range of topics, employment, agriculture, food security and prices, social assistance, remittances, migration, education, and health facility utilization. The team deliberately avoided asking questions directly related to illness within the household. Such questions were omitted for two reasons: first, to prevent non-response if households feared the authorities would come to remove ill members and, second, because the nature of the national sample was not well-suited to surveying disease incidences. The survey also included topics that were kept consistent in every round for monitoring purposes, such as those related to food security and economic activity, and some that were included in only one or two rounds based on the evolving situation. For example, the first round included questions as to whether the respondent had ever heard of Ebola and what sources of information they had on prevention. In later rounds, questions related to education were added, as schools reopened and social assistance as safety nets projects were rolled out.

The results from the survey yielded several important findings related to the economic situation. In both Sierra Leone and Liberia, the surveys found significant declines in employment during the crisis, but the effects were not significantly higher in places with higher numbers of Ebola cases. This indicates an overall economic slowdown caused by the nationwide precautionary measures, particularly the closure of markets, had more of an impact on employment than direct cases of Ebola. Moreover, in both countries, women were more likely to have stopped working during the crisis, and less likely to have returned to work by the end of data collection period. In Sierra Leone, income and labor force participation (hours) for both men and women remained below baseline levels at the end of data collection, although the overall percentage of individuals working had largely rebounded. In addition, many workers had switched sectors during the crisis, generally moving to positions with lower productivity (Fig. 3).

Beyond the findings related to labor markets, the surveys provided important insights related to prices, food security, coping strategies, education, avoidance of healthcare facilities, and perceptions of public safety and trust in institutions. The surveys were able to monitor the

2 Monitoring the Ebola Crisis Using Mobile Phone Surveys

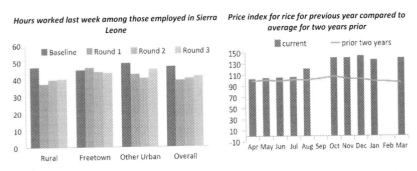

Fig. 3 Evidence from the Sierra Leone and Liberia phone surveys (*Source* Sierra Leone high-frequency mobile phone survey and Liberia high-frequency mobile phone survey)

usage of healthcare facilities for non-Ebola medical care. For example, the percentage of women in Sierra Leone giving birth in the previous two months in a hospital or clinic increased from 28% in November 2014 to 64% in February 2015 to 89% in May 2015. In some cases, these findings conflicted with the anecdotal evidence that had been previously guiding policy. In agriculture, farmers in both countries estimated that the production had declined, but to a lesser extent than had been feared, and with no evidence of the widescale abandonment that had been previously reported. In Sierra Leone, a delay in the arrival of seasonal rains also played a role. In education, once schools reopened, most students returned, 87% in Sierra Leone and 73% in Liberia. Of those that did not, the reason cited was monetary rather than fear of infection.

4 Implementation Challenges, Lessons Learned, and Next Steps

Although they cannot replace face-to-face household surveys in all contexts, mobile phone surveys offer substantial benefits in specific circumstances and for specific data collection needs. Advantages include the ability to collect data in volatile and high-risk environments (such as during political crises or epidemics), flexibility and responsivity to new data needs, timeliness, cost effectiveness, and utility for monitoring and impact evaluation. However, this approach remains challenging, and several lessons have been learned.

The risk of non-response and attrition applies to all panel surveys but is more likely for high-frequency mobile phone panel surveys. In the case of L2A, several strategies were undertaken to minimize these risks. Because sample selection did not consider prior ownership of a mobile phone, some households, particularly the poorest ones, had access to a mobile phone network but did not actually own mobile phones. To overcome this, mobile phones were distributed to all selected households, regardless of whether they already owned one, and respondents received training on various aspects of mobile phone ownership. In addition, the frequent power cuts in survey locations meant that phones could not be recharged, which could then lead to non-participation. To address these power cuts, small solar chargers were provided to allow households to charge their phones and receive follow-up calls.

In L2A, respondents were compensated each time they completed a phone interview, receiving a small amount of airtime credit transferred directly to their phones. This was both to compensate respondents for their participation, thereby encouraging them to stay involved, and to prevent the cancellation of phone numbers, which is a risk for those who do not 'top up' their phones after a certain period (usually 90 days). The lag period between the baseline survey and the first phone interview was also kept short. During the baseline survey, phone numbers were collected for all household members to increase the chances of reaching the respondent, and respondents were asked for their preferred call times. Efforts to track and trace hard-to-reach respondents also continued throughout implementation.

Response rates for the L2A surveys were generally high, reflecting the numerous measures taken to minimize non-response. In the Ebola surveys, however, other than providing limited compensation to respondents, it was not possible to take any of the above mitigation strategies. This was compounded by low network coverage rates, particularly in rural areas, and led to low response rates and issues with sample representativeness. For those baseline survey households that did not respond

2 Monitoring the Ebola Crisis Using Mobile Phone Surveys 23

in some of all of the cell phone rounds, analysts attempted to mitigate the impact of attrition by adjusting the weighting of the data. The correct weighting depends on whether cross-sectional or panel analysis is being conducted, and, in the case of panel analysis, which rounds of the survey are being compared. In the Sierra Leone and Liberia mobile phone surveys, multiple sets of weights were necessary depending on the combination of rounds. While the distribution of respondents in the mobile phone survey by age, gender, county, and sector of employment were similar to those found in the HIES and LFS samples, response rates were far lower in rural areas—compared with urban areas—due to limited network coverage. To adjust for differences in characteristics between the baseline and subsequent rounds, it was necessary to apply an attrition adjustment to the baseline survey weights. The adjustment included a propensity score adjustment, which uses the available characteristics of the household head from the baseline survey (age, gender, location, and sector of employment), and a post-stratification adjustment. This increased the total weighting of each stratum to match the distribution found in the last census. Full details of the weighting methodology can be found in World Bank (2014), and each report contains a table showing the regression results underlying the propensity score calculations on which the weighting adjustments were based. Even after taking into account these adjustments, however, careful review is necessary to determine if the results from the mobile phone survey can truly be considered representative, as opposed to merely indicative (Fig. 4).

Another lesson learned was to keep the survey short. While households can and will participate in a mobile phone interview, the questionnaire must be kept short to minimize respondent fatigue, which can be a cause of attrition and non-response. Mobile phone-based surveys are therefore not appropriate for lengthy interviews or complex questions, such as those relating to household consumption. Mobile phone surveys also cannot substitute in-depth information that can be collected in face-to-face household surveys.

While fielding new ad hoc surveys to monitor an evolving crisis is possible (see Chapter 3), a more systematic approach is clearly preferable. If a representative mobile phone survey could be carried out on short notice, this would not only provide valuable real-time

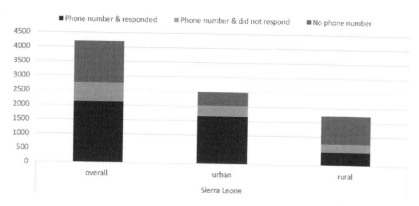

Fig. 4 Response rates for the high-frequency mobile phone surveys in Sierra Leone and Liberia (*Source* Authors' calculations)

information, but could also be used to mount an effective response. The high-frequency mobile phone surveys to monitor the socio-economic impacts of Ebola in Sierra Leone and Liberia were possible because the most recent national household surveys had collected contact information. A proactive approach to crisis monitoring would start with the systematic creation (and maintenance) of databases with phone numbers and core household respondent characteristics. Another lesson from the Ebola crisis is that setting up a call center is relatively straightforward and can even be done from abroad.

Annex 1: Links to Ebola Reports

Four reports were produced using the five rounds of the High-Frequency Cell Phone Survey on the Socio-Economic Impacts of Ebola in Liberia:

The socio-economic impacts of Ebola in Liberia: results from a high frequency cell phone survey (rounds one and two)—released in November 2014: http://documents.worldbank.org/curated/en/2014/11/24048037/socio-economic-impacts-ebola-liberia-results-high-frequency-cell-phone-survey.

2 Monitoring the Ebola Crisis Using Mobile Phone Surveys

The socio-economic impacts of Ebola in Liberia: results from a high frequency cell phone survey (round three)—released in January 2015: http://documents.worldbank.org/curated/en/2015/02/24051870/socio-economic-impacts-ebola-liberia-results-high-frequency-cell-phone-survey-round-three.

The socio-economic impacts of Ebola in Liberia: results from a high frequency cell phone survey (round four)—released in February 2015: http://documents.worldbank.org/curated/en/2015/02/24050332/socio-economic-impacts-ebola-liberia-results-high-frequency-cy-cell-phone-survey.

The socio-economic impacts of Ebola in Liberia: results from a high frequency cell phone survey (round five)—released in April 2015: http://documents.worldbank.org/curated/en/2015/05/24439139/socio-economic-impacts-ebola-liberia-results-high-frequency-cell-phone-survey-round-five.

Three reports were produced using the three rounds of the High-Frequency Cell Phone Survey on the Socio-Economic Impacts of Ebola in Sierra Leone:

The socio-economic impacts of Ebola in Sierra Leone: results from a high frequency cell phone survey (round one)—released in January 2015: http://www.worldbank.org/content/dam/Worldbank/document/Poverty%20documents/Socio-Economic%20Impacts%20of%20Ebola%20in%20Sierra%20Leone,%20Jan%2012%20(final).pdf.

The socio-economic impacts of Ebola in Sierra Leone: results from a high frequency cell phone survey (round two)—released in April 2015: http://www.worldbank.org/content/dam/Worldbank/document/Poverty%20documents/Socio-Economic%20Impacts%20of%20Ebola%20in%20Sierra%20Leone,%20April%2015%20(final).pdf.

The socio-economic impacts of Ebola in Sierra Leone: results from a high frequency cell phone survey (round three)—released in June 2015: http://documents.worldbank.org/curated/en/2015/06/24646532/socio-economic-impacts-ebola-sierra-leone-results-high-frequency-cy-cell-phone-survey-round-three.

Annex 2: Listening to Africa, Nutrition and Food Security Questionnaire

Today, we would like to ask you about food consumption in your household.

Nutrition

A1. In the past one week (7 days), how many days did you or others in your household consume any […]? IF NOT CONSUMED, PUT ZERO		NUMBER OF DAYS
A.	Cereals, Grains, and Cereal Products (Maize Grain/Flour; Green Maize; Rice; Finger Millet; Pearl Millet; Sorghum; Wheat Flour; Bread; Pasta; Other Cereal)	
B.	Roots, Tubers, and Plantains (Cassava Tuber/Flour; Sweet Potato; Irish Potato; Yam; Other Tuber/Plantain)	
C.	Nuts and Pulses (Bean; Pigeon Pea; Macadamia Nut; Groundnut; Ground Bean; Cow Pea; Other Nut/Pulse)	
D.	Vegetables (Onion; Cabbage; Wild Green Leaves; Tomato; Cucumber; Other Vegetables/Leaves)	
E.	Meat, Fish, and Animal Products (Egg; Dried/Fresh/Smoked Fish (Excluding Fish Sauce/Powder); Beef; Goat Meat; Pork; Poultry; Other Meat)	
F.	Fruits (Mango; Banana; Citrus; Pineapple; Papaya; Guava; Avocado; Apple; Other Fruit)	
G.	Cooked Foods from Vendors (Maize – boiled or roasted; Chips; Cassava – boiled; Eggs – boiled; Chicken; Meat; Fish; Doughnut; Samosa; Meal eaten at restaurant; Other cooked foods from vendors)	
H.	Milk and Milk Products (Fresh/Powdered/Soured Milk; Yogurt; Cheese; Other Milk Product – Excluding Margarine/Butter or Small Amounts of Milk for Tea/Coffee)	
I.	Fats/Oil (Cooking Oil; Butter; Margarine; Other Fat/Oil)	
J.	Sugar/Sugar Products/Honey (Sugar; Sugar Cane; Honey; Jam; Jelly; Sweets/Candy/Chocolate; Other Sugar Product)	
K.	Spices/Condiments (Salt; Spices; Yeast/Baking Powder; Tomato/Hot Sauce; Fish Powder/Sauce; Other Condiment)	

2 Monitoring the Ebola Crisis Using Mobile Phone Surveys   27

A1. In the past one week (7 days), how many days did you or others in your household consume any […]? IF NOT CONSUMED, PUT ZERO	NUMBER OF DAYS
L. Beverages (Tea; Coffee; Cocoa, Milo; Squash; Fruit juice; Freezes/Flavored Ice; Soft drinks such as Coca-Cola, Fanta, Sprite, etc.; Commercial Traditional-Style Beer; Bottled Water; Bottled/Canned Beer; Traditional beer; Wine or Commercial Liquor; Locally Brewed Liquor)	

Food Security

B1. In the past 7 days, did you worry that your household would not have enough food? Answer:

1 = Yes 2 = No

B2. In the past 7 days, how many days have you or someone in your household had to… IF NO DAYS, RECORD ZERO	DAYS
a. Rely on less preferred and/or less expensive foods?	
b. Limit portion size at meal-times?	
c. Reduce number of meals eaten in a day?	
d. Restrict consumption by adults in order for small children to eat?	
e. Borrow food, or rely on help from a friend or relative?	

B3. How many meals, including breakfast are taken per day in your household?	NUMBER
a. Adults	
b. Children (6-59 months) LEAVE BLANK IF NO CHILDREN	

B4. In the past "X" months [number of months since the last survey on this topic], have you been faced with a situation when you did not have enough food to feed the household? Answer: _____

1 = Yes 2 = No ≫B7

B5. When did you experience this incident in the last "X" months [number of months since the last survey on this topic]?

MARK X IN EACH MONTH OF 2016 WHEN THE HOUSEHOLD DID NOT HAVE ENOUGH FOOD.

LEAVE CELL BLANK FOR FUTURE MONTHS FROM INTERVIEW DATE OR MONTHS MORE THAN "X" MONTHS AGO FROM INTERVIEW DATE [number of months since the last survey on this topic].

2016											
Jan	Feb	Mar	Apr	May	June	July	Aug	Sep	Oct	Nov	Dec

B6. What was the cause of this situation? LIST UP TO 3 [Do not read options. Code from response].

CAUSE 1	CAUSE 2	CAUSE 3

Codes for B6:
1 = Inadequate household stocks due to drought/poor rains
2 = Inadequate household food stocks due to crop pest damage
3 = Inadequate household food stocks due to small land size
4 = Inadequate household food stocks due to lack of farm inputs
5 = Food in the market was very expensive
6 = Unable to reach the market due to high transportation costs
7 = No food in the market
8 = Floods/water logging
9 = Other (Specify): _____

B7. Does your household cope with food shortages in any of the following ways?		1 = Yes 2 = No
A.	Reduce number of meals eaten in a day	

2 Monitoring the Ebola Crisis Using Mobile Phone Surveys   29

B7. Does your household cope with food shortages in any of the following ways?	1 = Yes 2 = No
B.	Limit portion size at meal-times	
C.	Rely on less preferred and/or less expensive foods	
D.	Change food preparation	
E.	Borrow money, food, or rely on help from a friend or relative	
F.	Postpone buying tea/coffee or other household items?	
G.	Postpone paying for education (fees, books, etc.)?	
H.	Sell household property, livestock, etc.?	

B8. In case of food shortage, who eats less? Answer: _____
1 = Boys 0–15 years
2 = Girls 0–15 years
3 = Boys and Girls 0–15 years
4 = Men 16–65 years
5 = Women 16–65 years
6 = Men and women 16–65 years
7 = People over 65 years old
8 = Everyone eats equal amounts

References

Dabalen, Andrew, Alvin Etang, Johannes Hoogeveen, Elvis Mushi, Youdi Schipper, and Johannes von Engelhardt. (2016). *Mobile Phone Panel Surveys in Developing Countries: A Practical Guide for Microdata Collection.* Directions in Development. Washington, DC: World Bank. https://doi.org/10.1596/978-1-4648-0904-0. License: Creative Commons Attribution CC BY 3.0 IGO.

Demombynes, Gabriel, Paul Gubbins, and Alessandro Romeo. (2013). "Challenges and Opportunities of Mobile Phone-Based Data Collection: Evidence from South Sudan." Policy Research Working Paper 6321, World Bank, Washington, DC.

Himelein, Kristen, Mauro Testaverde, Abubakarr Turay, and Samuel Turay. (2015). *The Socio-Economic Impacts of Ebola in Sierra Leone.* Washington, DC: World Bank Group.

Hoogeveen, Johannes, Kevin Croke, Andrew Dabalen, Gabriel Demombynes, and Marcelo Giugale. (2014). "Collecting High Frequency Panel Data in Africa Using Mobile Phone Interviews." *Canadian Journal of Development Studies* 35 (1): 186–207.

World Bank. (2014). *The Socio Economic Impacts of Ebola in Liberia: Results from a High Frequency Cell Phone Survey (English)*. Washington, DC: World Bank Group. http://documents.worldbank.org/curated/en/428151468276303088/The-socio-economic-impacts-of-Ebola-in-Liberia-results-from-a-high-frequency-cell-phone-survey.

World Bank. (2016). "Socioeconomic Impact of Ebola Using Mobile Phone Survey in Guinea." World Bank Report No: ACS18659.

2 Monitoring the Ebola Crisis Using Mobile Phone Surveys

The opinions expressed in this chapter are those of the author(s) and do not necessarily reflect the views of the International Bank for Reconstruction and Development/The World Bank, its Board of Directors, or the countries they represent.

Open Access This chapter is licensed under the terms of the Creative Commons Attribution 3.0 IGO license (https://creativecommons.org/licenses/by/3.0/igo/), which permits use, sharing, adaptation, distribution and reproduction in any medium or format, as long as you give appropriate credit to the International Bank for Reconstruction and Development/The World Bank, provide a link to the Creative Commons license and indicate if changes were made.

Any dispute related to the use of the works of the International Bank for Reconstruction and Development/The World Bank that cannot be settled amicably shall be submitted to arbitration pursuant to the UNCITRAL rules. The use of the International Bank for Reconstruction and Development/The World Bank's name for any purpose other than for attribution, and the use of the International Bank for Reconstruction and Development/The World Bank's logo, shall be subject to a separate written license agreement between the International Bank for Reconstruction and Development/The World Bank and the user and is not authorized as part of this CC-IGO license. Note that the link provided above includes additional terms and conditions of the license.

The images or other third party material in this chapter are included in the chapter's Creative Commons license, unless indicated otherwise in a credit line to the material. If material is not included in the chapter's Creative Commons license and your intended use is not permitted by statutory regulation or exceeds the permitted use, you will need to obtain permission directly from the copyright holder.

3

Rapid Emergency Response Survey

Utz Pape

1 The Data Demand and Challenge

In 2017, the United Nations (UN) stated that the world was facing the most serious humanitarian crisis since the Second World War, with over 20 million people at risk of starvation and famine.[1] The crisis was concentrated in four countries: Nigeria, Somalia, South Sudan, and Yemen. Alongside hunger, large portions of the population in these countries were facing deteriorating living conditions and threatened livelihoods.[2]

---

[1]https://www.theguardian.com/world/2017/mar/11/world-faces-worst-humanitarian-crisis-since-1945-says-un-official.

[2]Food Security Outlook Update Nigeria, Famine Early Warning Systems Network (2017); Post-Gu Technical Release Somalia, Food Security and Nutrition Analysis Unit and Famine Early Warning Systems Network (2017); Food Security Outlook Update South Sudan, Famine Early Warning Systems Network (2017); Food Security Outlook Update Yemen, Famine Early Warning Systems Network (2017).

U. Pape (✉)
World Bank, Washington, DC, USA
e-mail: upape@worldbank.org

© International Bank for Reconstruction and Development/The World Bank 2020   **33**
J. Hoogeveen and U. Pape (eds.), *Data Collection in Fragile States*,
https://doi.org/10.1007/978-3-030-25120-8_3

The crisis was driven by both drought and conflict to differing degrees in the four countries. In Nigeria, the Boko Haram conflict contributed to poor market access, severe food shortages, and disruption of livelihoods in the North-Eastern States.[3] For the Somali population, the dry agricultural season contributed to high food prices, livestock losses, and displacement.[4] In South Sudan, below-average crop production and inter-communal violence contributed to famine in the former Unity State; in addition, 70% of the population of South Sudan was in serious need of humanitarian assistance.[5] In Yemen, airstrikes and violent clashes on the ground kept food prices high and resulted in high dependency on food imports and emergency aid.[6]

The crisis required a response along the humanitarian-development nexus, to address urgent humanitarian needs while working toward short- to medium-term socio-economic development goals. The UN and the World Bank have worked to synchronize their responses to crises to the greatest extent possible.[7] Greater development can improve resilience and reduce fragility, so that future shocks do not automatically lead to humanitarian catastrophes.[8]

During the crisis, rapid data collection was required to assess the population at risk of famine. Traditional survey methods were unsuitable for a variety of reasons. First, results were needed urgently, so lengthy household questionnaires were inappropriate. Second, funding constraints meant that costly traditional surveys were also unfeasible. Third, a significant portion of the affected populations was believed to be located in conflict-affected areas, where face-to-face data collection

---

[3]Food Security Outlook Update Nigeria, Famine Early Warning Systems Network (2017).

[4]Post-Gu Technical Release Somalia, Food Security and Nutrition Analysis Unit and Famine Early Warning Systems Network (2017).

[5]The UN officially declared famine in parts of Unity State, South Sudan: https://unmiss.unmissions.org/famine-declared-parts-south-sudan; Key IPC Findings: January–July 2017, Integrated Food Security Phase Classification (2017).

[6]Food Security Outlook Update Yemen, Famine Early Warning Systems Network (2017).

[7]Making the Links Work: How the humanitarian and development community can help ensure no one is left behind, Inter-Agency Standing Committee (2014).

[8]New Way of Working, United Nations Office for the Co-ordination of Humanitarian Affairs (2017).

is very risky. Given the context, there was a need for a survey that was low-cost, fast, and technically feasible. Data collection needed to be launched and completed in a matter of days, while also ensuring the safety of the implementing teams. Convincing sampling frames had to be obtained in environments where existing data was scarce. Finally, the crisis was unfolding in four different contexts, and country-specific approaches were required that were both standardized yet adaptive.

2 The Innovation

The Rapid Emergency Response Survey (RERS) was designed with standardized survey protocols that can be implemented quickly in times of crises. It was designed as a phone survey to allow rapid access to populations at risk of famine, and can be carried out by local call-centers at low cost.[9] During the crisis, enumerators recorded data digitally and uploaded it every day to a cloud-based server, in order to map and update data trends on a daily basis.

The questionnaire was quick to administer, yet still included a broad range of development topics that might need to be better understood during a crisis. A maximum administering time of about 20 minutes was necessary for many reasons: Phone networks often have weak connectivity, making long interviews difficult, respondents have shorter attention spans over the phone compared to face-to-face interviews, and minimizing respondent fatigue was crucial to increasing the accuracy of the data and to avoid burdening potentially stressed respondents. However, the questionnaire must also provide a wide snapshot of the population's conditions, investigating a comprehensive set of topics including education, livelihoods, health, market access, food security, and water, in order to identify which have been most affected by the crisis and to inform a response.

---

[9]The call centers were located in-country for Nigeria, South Sudan, and Somalia, and in Egypt for Yemen.

The survey covered the mobile phone users, with a focus on areas deemed to be in 'emergency' or worse by the Integrated Phase Classification (IPC).[10] In order to participate in the survey, households had to own a mobile phone, have network coverage, and a means to charge the phone. As such, one key limitation of mobile surveys is that it excludes households too poor to have a mobile phone, or households that are too isolated to live in areas with coverage. Despite this shortcoming, the survey allows for an immediate, ground-level assessment of challenges related to the crisis, and the survey's results can be considered conservative estimates of how the entire population is affected, leading to insightful policy interventions.

Sampling strategies must be adaptable to local contexts. In Nigeria, an ideal starting point was to call respondents from previous surveys who represented the intended population, since phone numbers had been collected for previous waves of this survey.[11] This approach allowed for the comparison of RERS estimates to estimates from the existing survey, which included the non-phone-using segment of the population. Household characteristics can thus be compared between the sample from the previous survey and the RERS, so that the representativeness could be assessed. About 80% of the phone numbers called resulted in successful interviews.

In the absence of existing surveys, a comprehensive list of phone numbers disaggregated by region would provide the best sampling frame; however, such lists are often unavailable, unreliable, or outdated. A bulk SMS to mobile phone users asking for consent to participate in a survey can provide an alternative sampling frame, from which respondents can be randomly selected. To ensure that the crisis-affected population is represented, it is crucial that any bulk SMS can geographically target crisis-affected regions: This approach was followed in Somalia. While this methodology is effective, it further compromises the representativeness of the survey by requiring respondents to reply to

---

[10]Guidelines on Key Parameters for IPC Famine Classification, Integrated Food Security Phase Classification (2016).

[11]General Household Survey 2016, conducted by National Bureau of Statistics of the Federal Government of Nigeria under Poverty and Conflict Monitoring Systems.

a text message before being interviewed. More than 65% of the numbers called resulted in successful interviews, allowing for fast execution of the survey. However, the actual number of recipients of the bulk SMS is unknown, making it difficult to calculate the percentage of SMS recipients who were interested in participating.

Random digit dialing (RDD) is a tool that randomly generates phone numbers, and can be a practical solution when phone number lists from surveys or bulk SMS dashboards are not available. This approach was followed in South Sudan and Yemen. Machine-learning algorithms can generate random-digit sequences based on a small set of verified existing numbers to create new numbers that are likely to exist. This reduces the loss of time that results from calling non-existent numbers. However, response rates are still unpredictable, especially if the survey targets specific geographic areas. On average, 10% of the numbers called resulted in successful interviews, prolonging the survey's duration.

3 Key Results

This section describes the collected data and highlights selected trends, starting with similarities between the four countries, and followed by selective country-specific deep-dives. In Nigeria, the survey involved households located in the North-East, North-Central, and South-South zones; of these, the North-East zone includes states classified to be under the emergency phase as per the IPC. For Somalia and South Sudan, only areas declared to be in a state of emergency or worse were surveyed. In Yemen, the survey covered all regions, stratified into emergency and non-emergency areas: Non-emergency regions are sampled because they had pockets of highly food-insecure households.

The proportion of highly food-insecure households was found to be large, but varied widely between the four countries, ranging from 30% in Somalia, to around 50% in Nigeria and Yemen, to over 70% in South Sudan (Fig. 1).[12] Higher food insecurity was recorded in

---

[12]Food security scores are based on the Reduced Coping Strategies Index Score and adapted to define lower scores for less food-secure households. Reduced Coping Strategies Index Score

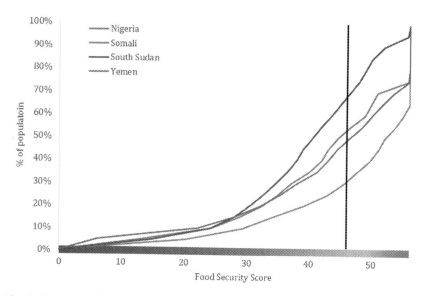

Fig. 1 Food security score by country (*Source* Author's calculations based on RERS data)

those countries that faced conflict during the crisis. A high incidence of conflict was reported in Nigeria, South Sudan, and Yemen, while in Somalia, the crisis was primarily due to dry agricultural seasons and a lack of resilience.

In addition to food insecurity, the populations surveyed also faced a range of developmental challenges. Livelihoods were affected in all four countries, with large portions of the populations (ranging from 31% in Nigeria to 84% in Yemen) facing a decrease in income and a change in their main source of livelihood (ranging from 13% in Nigeria to 31% in Somalia). Poor health, insufficient access to water, and low preparedness for drought were also common in all four countries. Other issues such as school attendance and livestock loss were more context-specific (Table 1).

---

is calculated using the CSI Field Methods Manual, Cooperative for Assistance and Relief Everywhere (2008).

Table 1 Developmental gaps among the famine-risk population

Food insecurity status	Nigeria[a]				Somalia				South Sudan			Yemen			
	All	Low	Medium	High	All	Low	Medium	High	All	Low	Medium-high[b]	All	Low	Medium	High
Number of households	581	173	96	312	2669	1290	556	823	1271	130	1141	1877	633	325	919
Percentage of households (%)	100	30	17	54	100	48	21	31	100	10	90	100	34	17	49
Change in income source (%)	13	13	11	14	31	26	27	40	21	22	21	26	24	24	29
Decrease in income (%)	31	21	25	38	44	36	39	60	42	33	43	84	76	85	90
Change in employment (%)	12	10	6	14	15	11	12	24	16	16	16	9	6	7	11
Not attending school regularly (%)	7	5	5	9	3	1	2	6	10	3	11	16	13	14	19
Didn't buy from market in 30 days (%)	12	20	14	8	10	4	7	22	17	28	16	12	12	8	14
Water isn't sufficient (%)					32	22	27	53	67	76	66	25	15	25	32
Household member sick (%)					47	36	46	65	91	80	92	48	34	51	57
No medication taken (%)					5	2	3	12	9	4	10	3	1	3	5
Low preparedness (%)					58	51	63	66	29	12	31	24	18	20	30

(continued)

Table 1 (continued)

Food insecurity status	Nigeria[a]				Somalia				South Sudan			Yemen			
	All	Low	Medium	High	All	Low	Medium	High	All	Low	Medium-high[b]	All	Low	Medium	High
Below average farm production (%)	10	5	10	13	15	14	19	15	27	25	27	15	14	14	16
Decrease in livestock (%)	10	10	5	12	26	25	24	28	25	15	26	14	12	12	16
Decrease in productive assets (%)	1	1	0	0	5	6	2	4	9	2	10	6	7	4	5

[a]Due to contextualization, data was not collected for certain topics marked in gray
[b]The 'medium' category had a very low sample size in South Sudan, leading to unreliable results. Thus, it has been combined with the 'high' category
Source Authors' calculations

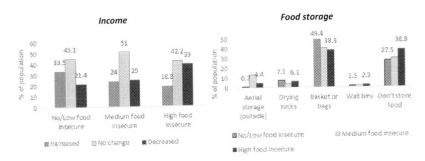

Fig. 2 Trends in income and food storage (*Source* Author's calculations based on RERS Nigeria data [Conducted with the National Bureau of Statistics of the Federal Government of Nigeria under Poverty and Conflict Monitoring Systems])

In Nigeria, one in five households lost income over the previous 12 months. Highly food-insecure households were more likely to experience a decrease in income than food-secure households (39 and 21% respectively; Fig. 2). Over one in three households did not store food for future use. Highly food-insecure households were most likely to not store food (39%) compared to households with low or medium food insecurity (28 and 29% respectively; Fig. 2). Early warning systems for drought preparedness and food-storage capabilities might allow for higher resilience and reduce the need for desperate coping strategies.

In Somalia, surveyed household had lost livestock and changed their employment activities. Over 30% of the Somali population owned livestock in the previous 12 months. However, among households that owned livestock, four in five faced a decrease in livestock holdings, with the primary reason being death or disease (66%). Livestock had also been depleted from being sold off (9% of livestock-owning households; Fig. 3). Economic assistance to compensate for livestock losses was clearly necessary to increase household income. The survey found that about 15% of households changed their main employment activity, with a shift away from farm-based employment (15–10% in farm labor; 20–13% in own-account farming), while the number of respondents involved in non-farm businesses increased sharply (17–28%; Fig. 3). Households responded to the drought by compensating for losses in agricultural income through shifts in the labor market.

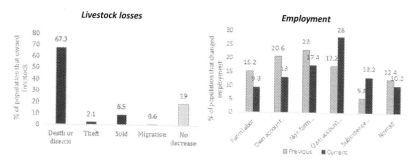

Fig. 3 Trends in livestock losses and employment (Somalia) (*Source* Author's calculations based on Somali RERS)

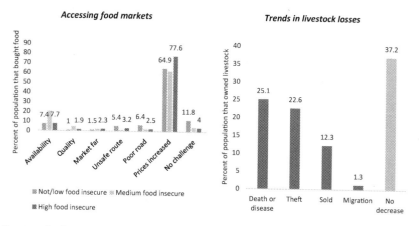

Fig. 4 Challenges in accessing food markets and livestock losses (South Sudan) (*Source* Author's calculations based on RERS South Sudan)

In South Sudan, most households rely on markets to buy food. However, while food is generally available in markets, households often cannot afford to buy food because of high prices (Fig. 4). These high food prices are not surprising, given South Sudan's recent period of high inflation. To improve access to food, the survey findings emphasize the importance of vouchers as opposed to food imports.

Almost 40% of the South Sudanese population owned livestock in the previous 12 months. Of these, almost two in three households lost livestock due to death or disease, and theft (25 and 21% respectively;

3 Rapid Emergency Response Survey

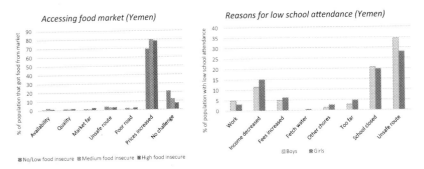

Fig. 5 Challenges in accessing food markets and reasons for low school attendance (Yemen) (*Source* Author's calculations based on RERS Yemen)

Fig. 4). Livestock restoration could help livestock-based agricultural livelihoods to be regained faster.

In Yemen, food prices and in particular, and school attendance were found to be affected by the crisis. Most households used markets to access food; as such, an increase in food prices was the key challenge in accessing the market for both food-secure and food-insecure households (Fig. 5). Again, this suggests that future interventions should be based on food vouchers rather than food imports. About one in four households had not sent all their children to school regularly in the previous year, largely due to unsafe routes to school and school closures. Safety issues and school closures resulted in low school attendance for both boys and girls (Fig. 5). This underlines the detrimental effects of insecurity on future generations, and the need to restore educational infrastructure.

4 Implementation Challenges, Lessons Learned, and Next Steps

The results of the Emergency Response Survey showcased above emphasize the importance of understanding the local situation, which is context- and country-specific. A quick turnaround from survey inception to

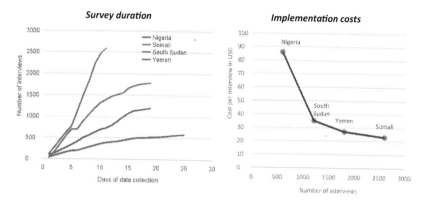

Fig. 6 Survey duration and implementation costs (*Source* Author's calculations based on RERS)

results and analysis is a key factor in the usefulness of the data to inform disaster responses. The survey was completed the quickest in Somalia (2600 interviews in ten days), and slowest in Nigeria (around 600 interviews in 25 days; Fig. 6), with the response rates and the size of the enumeration team being key factors in the speed of data collection. In Somalia, the enumeration team consisted of 25 enumerators, which was five times the size of the team in Nigeria. In South Sudan and Yemen, response rates were less than 15%, which increased the survey's duration. However, despite these various constraints, data collection was quick enough to generate results for each country in eight weeks. The case of Somalia further demonstrates that very rapid data collection can be done with a reasonably sized team of enumerators, even in a constrained environment.

This survey methodology can be deployed rapidly while keeping costs low. Operating through country-based call centers cost roughly $50,000 per country, and the cost per interview was less than $35 in all countries except for Nigeria. The interviews were most cost-effective in Somalia ($23 per interview) and most expensive in Nigeria ($86 per interview; Fig. 6). The bulk of the costs were fixed, and thus a larger sample size drove down the cost per interview. This fixed-cost structure allowed for increasingly cheap future rounds of the survey once the call center infrastructure has been established.

3 Rapid Emergency Response Survey   45

Statistical infrastructure, such as a list of respondents with phone numbers, can accelerate data collection and improve the representativeness of the survey. In Nigeria, phone numbers collected as part of a nationally representative household survey were used to select respondents. Such approaches can save time in preparing the sampling frame for the survey, compared to negotiating with mobile phone providers to provide randomized lists of phone numbers or to send a bulk SMS. It can also minimize the legal implications of mobile phone-based surveys, as some countries do not allow large numbers of unsolicited phone calls. Using data from a nationally representative survey, as was the case in Nigeria, the representativeness of the collected data can be assessed by comparing respondents who participated in the phone survey with respondents who could not be reached, as well as to respondents who either did not provide or did not have a phone number.

Questionnaire design is a crucial step in preparing a survey. In the absence of quantitative data about the impact of the drought on a variety of areas, the questionnaire was designed to explore diverse developmental topics including education, livelihoods, health, remittances, prices, and market access. Survey data clearly indicated that certain topics were more seriously affected than others, warranting more detailed exploration that was impossible in the context of the initial survey. For example, in South Sudan, more than 90% of households had suffered an illness in the previous three months (Table 1); however, while the questionnaire was able to collect details about the most recent illness in the household, the module was insufficiently deep to allow specific conclusions regarding health interventions to be drawn. In hindsight, additional information on household member-specific and less recent illnesses would have been valuable. However, the design of the questionnaire traded in-depth exploration against comprehensive thematic coverage. Another finding from South Sudan was that remittances were not severely affected by the crisis; again, in hindsight, the questionnaire could have been optimized by adding questions on remittances. However, it is impossible to make these choices a priori, especially during an unfolding crisis situation.

The use of an adaptive questionnaire is a promising approach to escape this limitation. The premise is to adapt the questionnaire while

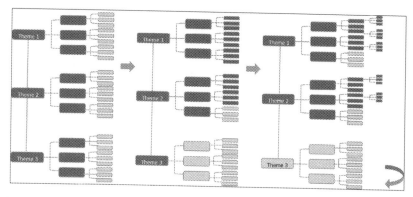

Fig. 7 Proposed adaptive questionnaire design (Color figure online)

collecting data. The first round of the questionnaire should cover a broad range of developmental topics, with an emphasis on preliminary questions assessing the extent to which each topic is affected. After around 500 interviews, trends from the data collection will indicate which topics warrant more exploration.[13] A survey conducted through a call center allows for the rapid adaptation of the questionnaire, as well as splitting the sample into individually representative parts at no extra cost. Thus, the questionnaire can be adapted after every 500 interviews, with increasing levels of detail on relevant topics and subtopics (Fig. 7, colored green). Less relevant topics can be dropped to keep the duration of the interview manageable (Fig. 7, colored gray). Even saving five minutes by skipping preliminary questions on irrelevant themes can save crucial time in a 20-minute interview for more in-depth exploration of relevant topics.

Adaptive questionnaire design fits well within the fast, low-cost survey methodology of the RERS. Enumerators can be trained on the full, detailed version of the questionnaire before data collection to allow quick adaptation of relevant and irrelevant topics (Fig. 8). The design will create systematically missing values for detailed questions in interviews conducted at the beginning of data collection, and for explorative

---

[13]Around 350 observations is the minimum sample size that provides a 95% confidence interval for estimates. Thus 500 is a sufficient sample size to map early data trends.

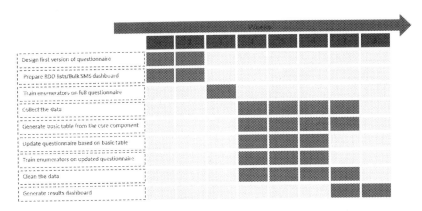

Fig. 8 Proposed timeline for adaptive questionnaire design with RERS methodology

questions later in the implementation. The random sequence of interviews, however, ensures that the missing data is not biased and, thus, can be analyzed by ignoring missing values. While this will affect standard errors, a large sample, such as the 2500 interviews in Somalia, can ensure sufficiently narrow confidence intervals, even after several adaptations of the questionnaire.

By presenting the example of a pilot survey, this chapter provides proof-of-concept that quantitative data collection via phone surveys is feasible, cost-effective, and informative in the context of shock responses. The pilot highlighted the importance of using an effective questionnaire design, like adaptive questionnaires, to balance the need to comprehensively cover a wide range of topics with the need to collect detailed information on specific sub-topics identified as part of the survey. Although smart and innovative designs can optimize trade-offs, emergency response surveys are no substitute for face-to-face household surveys based on representative sampling frames, ensuring that everybody is included, even the poorest and most vulnerable, who might not own mobile phones. However, compromises need to be made to provide a timely response following a shock, creating a niche for emergency response surveys as presented in this chapter.

Such emergency phone surveys can be prepared and implemented at global and national levels. At the global level, an adaptive questionnaire template can be prepared before emergency situations occur. This will reduce the preparation time needed to adapt the questionnaire template to a country and a specific crisis. At the country level, the groundwork for a survey can be prepared by collecting phone numbers of potential respondents. Questions about phone numbers and the willingness to participate in future survey interviews should be included by default in nationally representative surveys at both the household and firm levels. Lists of phone numbers of respondents who are knowledgeable about specific topics can further add value and allow for more in-depth interviews. These can be obtained by reaching out to sector ministries to collect phone numbers for their staff, which may already be integrated into the HR system. The ability to call, for example, health workers or police officers across an entire country allows for many more monitoring options that would not only be relevant in emergency situations. National statistics offices often maintain such (sufficiently anonymized) phone number databases, and provide them in emergency situations. In crisis-prone countries, the establishment of a call center and, potentially, the use of continuous phone surveys can also further accelerate implementation and provide baseline data.

Emergency phone surveys can play a critical role in crisis analytics, especially if integrated with other data sources that are either recurrently collected in a country, available upon demand, or typically collected during a crisis. For example, market price data is collected in most countries, whether by national statistics offices, UN agencies, or both, and can be triangulated geographically with household interviews. Satellite images can also provide additional context specifically to household interviews, for example, by gathering information about agricultural activity or damage to physical infrastructure. Furthermore, social network data or mobile phone usage data can provide invaluable insight. However, access to such datasets often takes time, making pre-emergency agreements necessary. UN agencies have developed standard data collection methodologies for crisis situations, including the WFP's Vulnerabilities Assessment and Mapping (VAM), and IOM's Displacement Tracking Matrix (DTM). For a meaningful integration, the different underlying data sources should be readily available and anonymized at the micro level, including sufficiently disaggregated geographical indicators. This requires pre-emergency agreements between involved agencies, ideally at both global and national levels. This is a valuable effort for the avoidance of redundancies and the creation of new synergies. Thus, technological advancements make crisis analytics an extremely powerful tool to inform crisis responses. To harvest their full potential, more pilots must be carried out, and specific statistical infrastructure as well as collaborations and agreements between different stakeholders will be needed.

The opinions expressed in this chapter are those of the author(s) and do not necessarily reflect the views of the International Bank for Reconstruction and Development/The World Bank, its Board of Directors, or the countries they represent.

Open Access This chapter is licensed under the terms of the Creative Commons Attribution 3.0 IGO license (https://creativecommons.org/licenses/by/3.0/igo/), which permits use, sharing, adaptation, distribution and reproduction in any medium or format, as long as you give appropriate credit to the International Bank for Reconstruction and Development/The World Bank, provide a link to the Creative Commons license and indicate if changes were made.

Any dispute related to the use of the works of the International Bank for Reconstruction and Development/The World Bank that cannot be settled amicably shall be submitted to arbitration pursuant to the UNCITRAL rules. The use of the International Bank for Reconstruction and Development/The World Bank's name for any purpose other than for attribution, and the use of the International Bank for Reconstruction and Development/The World Bank's logo, shall be subject to a separate written license agreement between the International Bank for Reconstruction and Development/The World Bank and the user and is not authorized as part of this CC-IGO license. Note that the link provided above includes additional terms and conditions of the license.

The images or other third party material in this chapter are included in the chapter's Creative Commons license, unless indicated otherwise in a credit line to the material. If material is not included in the chapter's Creative Commons license and your intended use is not permitted by statutory regulation or exceeds the permitted use, you will need to obtain permission directly from the copyright holder.

4

Tracking Displaced People in Mali

Alvin Etang and Johannes Hoogeveen

1 The Data Demand and Challenge

For decades prior to the 2012 rebellion, political leaders in northern Mali asserted that their people were marginalized and consequently impoverished. Separatist groups staged unsuccessful rebellions in 1990 and in 2007. In 2012, however, many of those fighting in the rebellion had received training from Gaddafi's Islamic Legion and were experienced with a variety of warfare techniques, and the rebellion that started with attacks on the Malian army in Menaka in mid-January 2012 culminated in a coup d'état by March 2012 and an attempt to take over the country by force. The three northern regions of Mali, Gao, Timbuktu, and Kidal became occupied by various rebel and Islamist factions until early 2013, when a coalition composed of the Malian

---

A. Etang · J. Hoogeveen (✉)
World Bank, Washington, DC, USA
e-mail: aetangndip@worldbank.org

J. Hoogeveen
e-mail: jhoogeveen@worldbank.org

© International Bank for Reconstruction and Development/The World Bank 2020
J. Hoogeveen and U. Pape (eds.), *Data Collection in Fragile States*,
https://doi.org/10.1007/978-3-030-25120-8_4

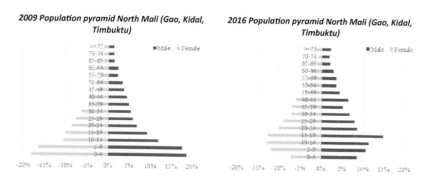

Fig. 1 Population pyramids before and after the 2012 crisis (*Source* Mali census data for 2009, INSTAT 2012; authors' calculations using January 2016 Permanent Monitoring baseline survey)

Army, French troops, and the ECOWAS-led International Support Missions to Mali (AFISMA) recaptured the occupied areas. Fighting between the Malian Army and the rebel factions broke out again in May 2014, and even though a peace accord was signed in June 2015, northern Mali remains insecure and contested.

At the height of the security crisis in Mali, over 500,000 people were displaced, nearly half of the estimated 1.2 million people who were living in the north (based on the 2009 population census). By October 2014, the number of displaced people was halved: the number of Internally Displaced Persons (IDPs) was estimated at 86,026, and the total number of Malian refugees was 143,471, with around 55,414 living in Mauritania, 53,491 in Niger, 32,771 in Burkina Faso, and 1330 in Algeria.[1,2]

The impact of the crisis on the population of northern Mali can be illustrated by looking at the age structure for the population in the north. Prior to the crisis, the population pyramid for the three northern regions was comparable to that of the entire country, but by 2015, the population pyramid for the north had changed considerably, reflecting the vast population movements that occurred during the crisis (Fig. 1). The biggest change occurred among children aged ten or younger.

---

[1] UNOCHA (November 2014): Mali: Evolution de Movements de Population.
[2] See UNHCR: http://data.unhcr.org/SahelSituation/country.php?id=501.

Information on the wellbeing of refugees and IDPs is typically hard to come by (Verwimp and Maystadt 2014), but is needed to formulate a response to the crisis. Information on returnees is particularly difficult to access. The reason for this is obvious: while it is relatively straightforward to interview people while they are displaced, tracking them after their return is much harder.

2 The Innovation

The Listening to Displaced People Survey (LDPS)[3] set out to address the information vacuum around the living conditions of displaced people and returnees. It did so in two ways. First, a baseline face-to-face survey was implemented that exclusively sampled displaced people, refugees, and returnees. Identifying the three target populations was made possible by the fact that each of these groups could be found in an identifiable location. Many displaced people were hosted by families in Bamako and had been registered by UN agencies; refugees were living in camps across the border, and returnees had returned to their locations of origin, predominantly in the northern cities of Gao, Kidal, and Timbuktu.[4] This approach to identifying returnees was possible because by August 2014, when the baseline survey was implemented, many displaced people had already started to return (see Fig. 2). The majority had returned between June and October 2013, a period that followed the signing of a peace deal between the interim government and the rebel factions to allow presidential elections to be held in July and August 2013.

During the baseline survey, information was collected on a range of household characteristics, including household composition, assets and

---

[3]Questionnaires, data and metadata of the LDPS are publicly available and can be downloaded from: http://bit.ly/2nsxSd6.

[4]It should be emphasized that locations were not randomly selected. Bamako was selected because it hosted a large number of IDPs, while the main cities in the north of Mali were chosen in order to obtain a large sample of returnees, given the available funds. A refugee camp in Niger was also chosen, as bureaucratic issues did not allow for the inclusion of a camp in Burkina Faso.

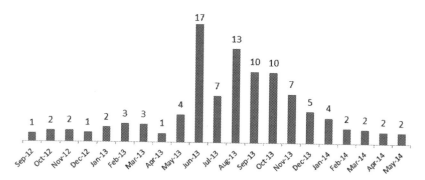

Fig. 2 Timing of return (percentage) (*Source* Authors' calculations based on the Mali Listening to Displaced People Survey)

income sources, as well as food security and experiences during the crisis. The baseline survey also asked perception questions about trust, security, about changes in wellbeing and perspectives on the future.

To track living conditions over time, the baseline survey was complemented with follow-up mobile phone interviews. This approach had the added advantage that if households chose to return during the research period, they remained within the sample. The ability to trace displaced people while they were still on the move was the most important innovation of the LDPS.

The baseline survey was used to identify respondents for the mobile phone interview. Because the survey intended to ask questions about perceptions and was seeking to be representative of the adult population, it was important that one adult was identified from within each household to be the main respondent throughout the survey period. It was equally important for the sake of representation that the person was not always the head of household. As a result, within each household, one person was selected randomly from all household members above the age of 18. Respondents were equally split between men and women to obtain a good representation of the opinions of both genders.

Upon completion of the baseline interview, all respondents received a mobile phone to avoid bias with regard to phone ownership. Mobile interviews were conducted in monthly intervals, using a specialized call center in Bamako. Interviews were conducted in the relevant local languages, French, Bambara, Kel-Tamashek, or Songhai. During the phone interviews (lasting 20–30 minutes) structured questions were asked about the welfare of the household and changes in location, as well as perception questions. Upon completion of the interview, respondents received a small token of appreciation in the form of US$2 worth of phone credit.

Over a period of twelve months, from August 2014 to August 2015, monthly interviews were conducted. The original sample comprised 501 respondents (51% men, 49% women) split between IDPs located in the capital city of Bamako (100), refugees living in refugee camps in Mauritania (100) and Niger (81), and returnees living in northern Mali, in the regional capitals of Gao (90), Timbuktu (80), and Kidal (50).

3 Key Results

The households in the sample only comprise displaced or formerly displaced people, so to investigate how those in the sample compare to non-displaced households, they need to be compared with existing data. Figure 3 illustrates the comparison for level of education, against baseline data collected prior to the crisis in 2011. It compares levels

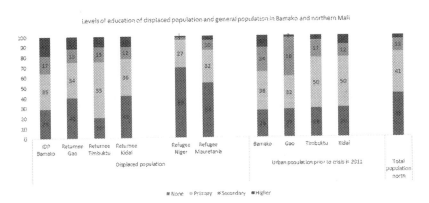

Fig. 3 Level of education of population aged 18+ (percentages) (*Source* Authors' calculations using the Listening to Displaced People Survey and the Enquête Modulaire et Permanente, EMOP 2011, of the Mali National Institute for Statistics, INSTAT)

of education of adults in the four cities of Bamako, Gao, Timbuktu, and Kidal. It is important to note that levels of education in Mali are extremely low. Even in the capital city of Bamako, more than half of the adults have not progressed beyond primary education, while in Kidal and Timbuktu, 80% completed primary education at most. In comparison, IDPs and returnees are better educated, aside from those in Gao. IDPs in Bamako have levels of education comparable to the general adult population of Bamako, which is higher than that of the urban population in the north. Returnees are also more likely than the overall populations of Kidal and Timbuktu to have achieved secondary education or higher.[5]

Refugees, in contrast, are less educated. In particular, refugees who went to Niger have lower levels of education than the overall population of northern Mali.

Regarding consumer durables, all three sub-populations, IDPs, refugees, and returnees were revealed to have higher levels of ownership than the average citizen of the north. As such, despite the loss of consumer durables due to the crisis, IDPs, refugees, and returnees still own more than or similar amounts of assets to the average population of the north prior to the crisis. This is shown in Fig. 4, which presents the proportion of IDPs, refugees, and returnees who own assets after the crisis and compares this with the percentage of households who owned assets prior to the 2011 crisis in Gao, Timbuktu, and Kidal. The value of assets owned by IDPs and refugees was found to be comparable to that of households between the third and fourth wealth quintiles, locating displaced peopled in the middle or upper-middle classes. As with education, displaced people's levels of asset ownership are more comparable to those of the average citizen in Bamako rather than the average citizen of the urban areas of Gao, Timbuktu, and Kidal.

This finding that displaced people were better off than others is confirmed by Peña-Vasquez and Mueller (2017), who use the same database. They conclude that people were more likely to opt for displacement when they felt more at risk, when they were relatively better

---

[5]Some of the results presented in this section have also been reported in Etang Ndip et al. (2016).

4 Tracking Displaced People in Mali

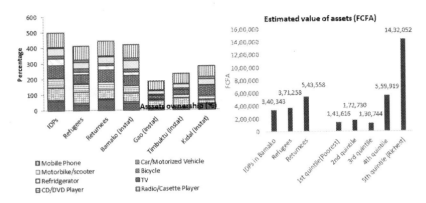

Fig. 4 Asset ownership compared with regional average (*Source* Authors' calculations using the Listening to Displaced People Survey, 2014 and the Enquête Modulaire et Permanente, EMOP, 2011 of the Mali Institute of Statistics (INSTAT))

off, and interestingly, when they lived in villages with greater access to transportation, either by land or water.

The main purpose in tracking displaced people, for the purposes of this chapter, is what the survey can tell us about their living conditions over time. The results show how the respondents' perception of their living conditions changed over time and across locations. In wave 12 in Kidal, for instance, there is a large decrease in the proportion of respondents stating that their living conditions were worsening, and an increase in respondents stating that they remained the same. This wave followed the signing of the Peace Accord in June 2015; however, the optimism found in Kidal at this time was not shared by the other three cities covered by the survey (Fig. 5).

The data collected takes the form of a longitudinal (panel) dataset, which allows to control for individual fixed effects. Hoogeveen et al. (2019) exploit the panel nature of the dataset to investigate the drivers of the decision to return, exploring how employment status, security, and expectations affect people's willingness to go back home. The findings suggest that the decision to return is affected by a comparison of (opportunity) costs and benefits, but also by other factors: Individuals who are employed while displaced are less willing to return home, as are better-educated individuals, or those receiving assistance. The

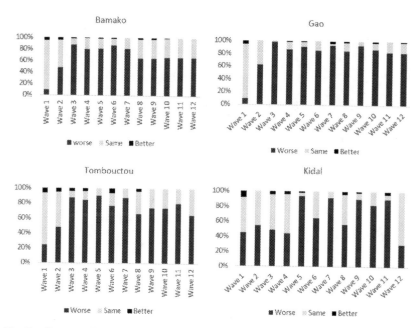

Fig. 5 Changes in perceived living conditions over the duration of the survey (*Source* Authors' calculations based on Mali Listening to Displaced People Survey)

opposite is true for ethnic Songhais and people from Kidal. The results show that individuals with higher levels of education do better when displaced, and if they return, they find jobs more easily than those with less education.

Using all twelve waves of the survey, Hoogeveen et al. ran a fixed effects linear probability model. These individual fixed effects capture all time-invariant individual characteristics such as ability, education, and stamina, as well as several stable household characteristics and environmental factors (e.g. attitude toward refugees or IDPs in the local community), while the time fixed effects control for events specific to a time period, such as weather shocks or military events. They find that those who found employment while being displaced were significantly less likely to return, while refugees and those who owned a gun were more likely to return (Fig. 6).

Those who found employment while being displaced were less likely to return to northern Mali; refugees and those who owned a gun were more likely to return.

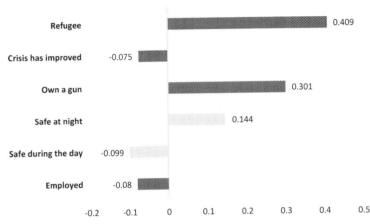

Fig. 6 Fixed effects regression on the decision to return (*Source* Hoogeveen et al. 2019)

4 Implementation Challenges, Lessons Learned, and Next Steps

The success of the tracking survey depended on the ability to maintain a stable sample. The measures employed were not unlike those discussed in Chapter 2: respondents received phones, were rewarded for participation with phone credit, and were given the opportunity to carry out the interview in their own language. The survey team emphasized approaches that might reduce drop-out, e.g. respondents were asked to indicate the time at which they preferred to be called. During the call, they would always speak to the same enumerator, thus building rapport. In the refugee camp in Mauritania, response rates declined due to weak network coverage. This was resolved by working with field-based enumerators who relayed the responses back to the call center in Bamako. The team also asked community members to follow up on respondents who could not be reached over the phone. This tracking mechanism was set-up at the survey design stage by collecting alternative phone numbers of the respondents such as phone numbers of other household members, friends, and neighbors. This helped enumerators reach respondents who did not answer their own phones. These measures were effective: the non-response rate was very low, between 1 and 2% per wave. The percentage of households not responding to more than two consecutive rounds, was even lower, only 0.8%. Attrition rates bore little relation to the movement of the respondent. For instance, in the area with the highest amount of movement, Bamako, the initial sample comprised 100 households. Of these, 12% indicated one year later that they had moved, but only one household dropped out of the sample. A similar finding holds true for Gao, where the sample initially comprised 90 households, and although some 7% moved, only two households dropped out of the sample.

Not only is the stability of the sample quite remarkable, but this survey also demonstrates that mobile phone surveys are useful tools for collecting data in hard-to-reach places. The case of Kidal, a desert town, illustrates this point. Kidal lies in a remote corner of northern Mali and is only accessible by 'piste' (i.e. unmarked dirt road), and the nearest town, Gao, is 285 km away. Moreover, during the period in which the data were collected, the government of Mali exercised no control over the town. Despite these factors which would normally greatly hinder data collection, the mobile phone survey collected information on a monthly basis with response rates that were near-universal (see Fig. 7, right panel).

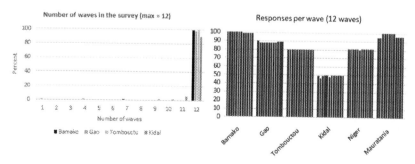

Fig. 7 Attrition rates (*Source* Authors' calculations using the Mali Listening to Displaced People Survey)

The ability to follow respondents as they change locations offers exciting new possibilities for welfare monitoring, as movement is often associated with large societal changes in welfare. We know, for instance, that rural-to-urban migration is associated with declining poverty of the movers in a process called structural transformation, in which increases in agricultural production facilitate rural–urban migration by increasing rural incomes while simultaneously suppressing (urban) food prices. Once this process starts, markets become more important, the non-farm and agribusiness sectors grow, and the food value chain and rural–urban linkages are strengthened. As rural incomes grow even further, second-order effects emerge: the stock of human and physical capital increases as households invest part of their increased incomes in their offspring. This leads to further productivity gains, and to emigration of better-educated people. While this process is well-understood, surprisingly little is known about how individual migrants fare during the process of transition. Nor is much understood about the characteristics of successful migration, as opposed to migration in which one ends up chronically poor in an urban slum. Mobile phone tracking surveys can be used to collect the data needed to fill this knowledge gap, and can be applied equally to returning IDPs and refugees, to school leavers, to those completing a job training program, or those having gone through a DDR program.

References

Etang, Alvin, Johannes Hoogeveen, and Julia Lendorfer. (2016). "Socioeconomic Impact of the Crisis in North Mali on Displaced People." *Journal of Refugee Studies* 29 (3): 315–340.

Hoogeveen, J., M. Rossi, and D. Sansone. (2019). "Leaving, Staying, or Coming Back? Migration Decisions During the Northern Mali Conflict." *The Journal of Development Studies* 55 (10): 2089–2105.

INSTAT. (2012). 4ème Recensement Général de la Population et de l'Habitat du Mali (RGPH-2009). Analyse des résultats définitifs. Thème: scolarisation, instruction et alphabétisation au Mali.

Peña-Vasquez, A., and D. Mueller. (2017). Consequences of Conflict: Forced Displacement, Insecurity, and Transportation in Northern Mali. Mimeo.

Verwimp, P., and J.-F. Maystadt. (2014). "Forced Displacement and Refugees in Sub-Saharan Africa. An Economic Enquiry." Policy Research Working Paper no. WPS 7517, World Bank Group, Washington, DC.

The opinions expressed in this chapter are those of the author(s) and do not necessarily reflect the views of the International Bank for Reconstruction and Development/The World Bank, its Board of Directors, or the countries they represent.

Open Access This chapter is licensed under the terms of the Creative Commons Attribution 3.0 IGO license (https://creativecommons.org/licenses/by/3.0/igo/), which permits use, sharing, adaptation, distribution and reproduction in any medium or format, as long as you give appropriate credit to the International Bank for Reconstruction and Development/The World Bank, provide a link to the Creative Commons license and indicate if changes were made.

Any dispute related to the use of the works of the International Bank for Reconstruction and Development/The World Bank that cannot be settled amicably shall be submitted to arbitration pursuant to the UNCITRAL rules. The use of the International Bank for Reconstruction and Development/The World Bank's name for any purpose other than for attribution, and the use of the International Bank for Reconstruction and Development/The World Bank's logo, shall be subject to a separate written license agreement between the International Bank for Reconstruction and Development/The World Bank and the user and is not authorized as part of this CC-IGO license. Note that the link provided above includes additional terms and conditions of the license.

The images or other third party material in this chapter are included in the chapter's Creative Commons license, unless indicated otherwise in a credit line to the material. If material is not included in the chapter's Creative Commons license and your intended use is not permitted by statutory regulation or exceeds the permitted use, you will need to obtain permission directly from the copyright holder.

5

Resident Enumerators for Continuous Monitoring

Andre-Marie Taptué and Johannes Hoogeveen

1 The Data Collection Challenge

The conflict in Mali in 2012 broke out after a long period of political and economic stability. It began when armed separatist groups occupied the northern desert and semi-desert regions. A period of instability followed, during which an estimated 36% of the total population from the affected regions fled to the south of Mali and to neighboring countries. The crisis had dramatic effects on public infrastructure and service and reduced people's mobility and their access to markets. It also led to the destruction and theft of assets and shook investor confidence. Farmers were cut off from their fields, artisans were unable to sell their produce as tourism came to a halt, traders were unable to move, and breeders

---

A.-M. Taptué (✉) · J. Hoogeveen
World Bank, Washington, DC, USA
e-mail: ataptue@worldbank.org

J. Hoogeveen
e-mail: jhoogeveen@worldbank.org

© International Bank for Reconstruction and Development/The World Bank 2020
J. Hoogeveen and U. Pape (eds.), *Data Collection in Fragile States*,
https://doi.org/10.1007/978-3-030-25120-8_5

with high numbers of livestock were forced to leave conflict-affected areas for safer places, losing many of their animals to theft along the way. The crisis reinforced the feeling of neglect by the Malian state among those from the affected areas, while simultaneously strengthening cross-border ethnic loyalties and economic ties. The conflict officially ended with the Peace Accords signed in May and June 2015, but the North remains insecure, as it has become a safe-haven for terrorists and criminals.

The crisis created distrust between different ethnic groups and among people of different religious affinities. Social cohesion weakened, and interactions became more restricted, inducing a feeling of fear. About one in three living in Timbuktu or Gao reported in July 2016 that they did not feel safe at home at night; in Kidal, this rose to two in three. Many people distanced themselves from social networks, neighbors became estranged, mixed marriages ended, and even within families, members became wary of each other. Animosity was also expressed toward the government. By July 2016, 53% of the population had lost confidence in the government, and confidence in the judicial system ranged from 66% in Timbuktu to as low as 8% in Kidal.

Collecting data in these circumstances is challenging, especially for emissaries of the central government. In fact, since the outbreak of the conflict in 2012, agents of the National Institute of Statistics have been unable to collect any data in Kidal or elsewhere in northern Mali. Data, however, was urgently needed to monitor the developments post-signing of the Peace Agreement. The Peace Agreement had established conditions for the restoration of stability and economic recovery, and called for development planning and new investments in the north, as well as the creation of a monitoring system to assess the impact of assistance on security, socio-economic development, and wellbeing.

The Permanent Monitoring System (PMS) was created to respond to this data challenge. It consists of an observatory that relies on local enumerators living in northern Mali, who collect data on a monthly basis. The PMS amasses information from a representative sample of

5 Resident Enumerators for Continuous Monitoring   65

households living in the targeted areas,[1] and from local authorities, clinics, schools, and markets, where commodity prices are collected.[2]

2   The Innovation

When enumerators from outside a community are not welcome or when travel to and within a region is dangerous for outsiders, one solution is to work with locally recruited enumerators who reside in the area. The use of such 'resident' enumerators is usually discouraged, particularly for consumption surveys as experience shows that due to limited possibilities for supervision, data quality tends to erode over time, while respondents tend to grow tired of answering detailed questions repeatedly about the various items they have consumed. For this reason, many consumption surveys have shifted from collecting consumption data using diary-methods to approaches that rely on recall. In the former approach it is often necessary for enumerators to stay in the village for up to a month; using recall methods survey teams can stay in the village for much shorter periods of time.

---

[1]As many households had fled the area, old sampling frames were no longer valid. To assure a representative sample was none the less collected each enumerator had a list of local landmarks as well as a direction to move. Enumerators would start at the landmark and sum the date of the day till one digit was obtained. That was the number of the first household to be interviewed counting from the landmark. Subsequently every second (rural areas) or every fifth household was interviewed till a total of 5 households was interviewed after which they would move to the next landmark. The selection of individuals within the household to answer the questionnaire was conducted as follows: the head of household (male or female) was selected to answer the first part of the questionnaire dealing with general questions about the households. Using the roster of household members which was compiled during the first part of the interview, another member of the household aged 18 or above was selected randomly to answer the second part of the questionnaire in which perception questions were asked. Alternation between male and female was ensured. The survey thus generated data that are reflective of the opinions of those aged 18 and above in northern Mali.

[2]All data are made publicly available (http://www.gisse.org/pages/miec/suivi-permanent1.html), and reports have been widely disseminated.

There are major advantages to an approach that relies on enumerators that reside for longer periods in a village. Among these are that resident, locally recruited enumerators know the survey areas well. This reduces many of the complexities associated with insecurity, local grievances, or language. The latter is a critical advantage. Ethnolinguistic fractionalization is high in Africa and in many locations, the ability to speak the language of choice of the respondent is key to the success of a survey. When enumerators cannot phrase questions in the language that respondents are most comfortable with, responses may be wrong or biased.

Another advantage of using resident enumerators is that, contrary to survey teams that visit an enumeration area for a short period of time, resident enumerators have ample time to carry out interviews. Capitalizing on this, enumerators of the PMS were asked to administer five different survey instruments including a household survey that contained multiple modules among which socio-demographic characteristics and income-generating activities, including detailed questions on agriculture, livestock, fisheries, and entrepreneurial activities. The survey collected information on assistance received, the return of refugees and internally displaced people, the health of household members, and food security. The household survey also covered shocks that households might have experienced, possession of assets, and access to services. A part of the questionnaire was devoted to subjective questions about the implementation of the Peace Agreement, perceptions of security, and priorities for initiatives that could consolidate peace and security in the region.

A second survey instrument was used to interview local authorities (mayors, traditional authorities, and local chiefs) to collect information about local initiatives and interventions, as well as information about the evolving security situation. A third survey was administered to assess the operations of health care centers in the surveyed villages. This survey assessed the impact of the crisis on their functioning, the presence and return of staff that had fled during the conflict, the assistance they received, and their needs in terms of supplies and equipment. A fourth survey was conducted in primary schools in the surveyed villages. Like the clinic survey, it assessed the presence of staff

5 Resident Enumerators for Continuous Monitoring 67

and the return of teachers who had fled during the crisis, the assistance the schools had received, and the school's needs. The fifth and final survey instrument collected information on prices of a selected list of commodities to gauge the changes in the cost of living in different localities.

Another advantage of resident enumerators is that they are in a better position to deal with 'moving' populations such as herders, of which there are many in northern Mali. Herders move about, few have mobile phones, and even if they do possess them, they are often out of range of a telephone network. This means that phone interviews as discussed in Chapters 2 and 3 are not feasible, particularly if non-random non-response is to be avoided. The ability to deal with moving respondents is determined by the time availability and local knowledge of resident enumerators. Locally recruited enumerators know where to find pastoralists as they regularly gather at specific locations to water their animals, or as they move from pasture to pasture following well-established grazing patterns. Herders are not the only mobile population. In many places in Africa, farmers also move. During the growing season, many remain at their fields in temporary shelters only to return to their village after the harvest. Enumerators selected from the village can follow households to their farms for interviews. They know the area and, unlike survey teams visiting enumeration areas for only a short period, have more flexibility in when to carry out an interview. They can meet respondents in the evening, early in the morning, at the market, or at the place where the respondent carries out his or her business.

Once enumerators have developed good relationships with the respondents, and respondents have confidence in them, resident enumerators are more likely to elicit accurate information, particularly when questions are sensitive, and the enumerator is able to emulate that responses will be kept confidential. This is another advantage of resident enumerators: it dispels fear and creates trust, trust that can be difficult to establish between people from different localities or ethnic groups. The importance of this cannot be underestimated in a (post)conflict situation, as it is not uncommon for respondents in insecure locations to fear reprisals for having provided information to an outsider, no matter how innocuous the information may seem. However, if the enumerator stays with the respondents in the village, it signals trustworthiness and allays such fears.

The flexibility and level of trust the locally recruited enumerators built in northern Mali allowed them to collect high-quality data and to assure high response rates over the course of a year. Between January 2016 and January 2017, the highest household non-response rate encountered was 4.4% in October 2016 in Gao, when ten households did not respond to the survey; however, they resumed their participation in January 2017. During this particular month, the northern regions experienced 21 attacks and bomb explosions, including six in Gao, four in Timbuktu, and two in Kidal. Still these insecurity events did not disrupt the survey or cause the response rate to drop.

There were two clear challenges when it came to relying on resident enumerators. The first is that it may be difficult to identify skilled enumerators in the communities of interest. Particularly in remote locations, the number of suitable candidates may be limited. Few people are likely to have experience with survey data collection and finding people with a certain level of formal education may be difficult. In the case of the northern Mali survey, the pool of eligible candidates was further reduced by the requirement that prospective enumerators had their own means of transport, allowing them to move about easily, while at the same time, assuring a greater sense of ownership and responsibility than one might expect for a project provided means of transport. A second challenge is supervision.

Hiring and managing enumerators was delegated to a local firm with extensive experience in data collection, and with a robust network in northern Mali.[3] It advertised the positions on its website and mobilized its contacts in the region to publicize the job opportunity. To assure the independence and objectivity of enumerators, the firm avoided relying on local authorities for recruitment. Those applying were expected to send proof of their education (diplomas) and of the

---

[3]Different contexts and data demands call for different solutions to this problem. E.g. in the context of a national public works program in the Central African Republic (LONDO project), the team used former locally recruited team leaders to collect information on how beneficiaries used the bicycles they had received after the project had left the area. Their reason to rely on these former employees was that the areas would be difficult of access by survey teams, while security costs would be prohibitive. Moreover, former team leaders had deep local knowledge and were able to find the beneficiaries through local social networks as many have no phones.

5 Resident Enumerators for Continuous Monitoring 69

possession of a motorbike (as a means of transport); these were later checked during enumerator training and the first supervision mission. Enumerators were informed that they would not be engaged full-time, and could take up other employment as long as they were available for this activity for at least one year, and during the first two weeks of each month.

Enumerators who had finished at least secondary education were sought, but it proved challenging to find people with that level of education living in remote villages, as less than 5% of the population of northern Mali has completed secondary education or higher (INSTAT 2012). In the end, and after devoting much effort to identifying and hiring enumerators in each enumeration area, it was not possible to find sufficiently qualified people for each location. As a solution, certain enumerators were required to cover two or three villages close to one another, a solution that caused few problems since enumerators had their own motorbikes to move between villages.

For those who did qualify, wages were high. To complete some 20 questionnaires over a two-week period every month, enumerators were paid approximately US$350 per month, plus a premium of US$600 every quarter and again at the end of the operation. These premiums were needed to assure the continued participation of the enumerators, as other organizations active in the area were offering competitive salaries. Although the budget for enumerator fees was relatively high, the overall cost of one round of data collection was reasonable: about US$30,000 for approximately 800 questionnaires (12 households per enumeration area, plus school, clinic, district leaders, and price questionnaires), or less than US$40 per questionnaire. The reason for this relatively low unit cost, despite high salaries is due to the minimal expenses incurred for transport, printing and communications, and meals and lodging (Fig. 1).

In the end, 35 enumerators were hired. Since traveling to northern Mali was not recommended for people not from the area, all enumerators were invited to Bamako for one week of training. In addition to becoming familiar with the survey material, questionnaires and manuals, much emphasis was placed on how to behave, as the aim was for

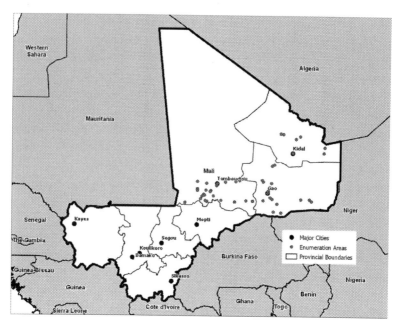

Fig. 1 Map indicating the location of enumerators across northern Mali (*Source* World Bank 2016)

the enumerators and respondents to develop an ongoing relationship for a period of more than one year. Hence, the training emphasized the importance of confidentiality, the importance of maintaining good relationships with respondents and local authorities, and the necessity of remaining neutral when collecting data.

Maintaining data quality was not an issue, as the response rates presented in Table 1 illustrate. Not only were enumerators motivated, as demonstrated by the fact that none dropped out of the exercise, but the use of tablets to collect data and the ability to remotely supervise the enumerators' actions improved data quality dramatically. The tablets registered the data and the time of data collection, along with the GPS coordinates of where the data were entered. This para-data allowed the firm supervising the data collection to assess whether the enumerators had indeed visited households for interviews, and to assess the average response time. The use of Computer-Assisted Personal Interviewing

5 Resident Enumerators for Continuous Monitoring 71

Table 1 Response rate (percentage of households that answered the survey)

		2016								2017
Region	Obs.	Jan	Feb	Mar	Apr	May	June	July	Oct	Jan
Gao	227	100	100	100	100	100	100	100	95.6	100
Kidal	121	100	100	100	100	100	100	98.2	98.3	98.3
Timbuktu	324	100	99.7	100	100	100	100	100	100	100

Source Authors' calculation using data from the Permanent Monitoring System

(CAPI) thus solved an important supervision problem that might otherwise have affected the quality of data collected by local enumerators operating under limited supervision.

To facilitate data collection using tablets, enumerators were trained in the use of CAPI techniques. Different questionnaires (for households, schools, clinics, and district leaders, and price questionnaires) were programmed in CSPRO, and a server was installed in the office of the firm supervising the work. Using CAPI allowed enumerators to send data to Bamako as soon as they completed a survey and had access to the internet. Though phone network coverage is limited in northern Mali, the network exists, at least in the urban center of each district. It was agreed that at least once a week, enumerators would move to a location that had network coverage to transfer their data to the server. At the beginning of each month, when enumerators were within reach of the phone network, they were paid using mobile payment systems such as Orange Money. Enumerators also downloaded new or updated questionnaires at these times. Relying on CAPI thus allowed the team to dynamically change the questionnaires used. Core questions typically did not change, but the questionnaires were adapted regularly to respond to new requests for information from development agencies and the government. Questionnaires were also changed in response to events on the ground and enumerators were expected to report noteworthy events, the distribution of material to farmers and breeders, and the functioning of schools and clinics. Their feedback was then used to update the questionnaires.

CAPI was not used everywhere: in some villages in Kidal, paper questionnaires continued to be used as respondents had expressed concerns about the use of tablets. They feared enumerators might use the GPS capacity of the tablet to order drone strikes. In these few instances, enumerators filled in paper questionnaires and subsequently transferred the responses onto the CAPI system, before electronically sending the responses to the server in Bamako.

The firm visited each enumeration area every six months for additional supervision, exposing the supervision team to insecurity while traveling, but once in the villages, the team was generally given a warm welcome. The team would meet with local authorities, including traditional and religious authorities, to (re)explain the objectives of the activity, and to request continued collaboration. The team also met with citizens at large, stressing how the enumerators were working in the interests of the whole community by striving to collect good information on the issues affecting their villages.

These efforts were successful. Quality data was collected throughout the entire period and for more than one year, the PMS informed the government of Mali and international organizations on changes in the situation in northern Mali. Best of all, none of the enumerators were harmed, nor was any survey respondent affected by violence that could in any way be associated to the survey.

3 Key Results

From September 2015 to January 2016, 35 enumerators covered 672 households across 56 villages and city areas, administering the five different types of survey instruments. Some key results are presented below. Food insecurity was found to mostly affect households in Gao and particularly in the early months of the year, when more than one-quarter of Gao's citizens lived in a state of food insecurity; this declined to around one-fifth between March and October. In Kidal, food insecurity was found to be much less of an issue, with less than 10% of households living in a state of food insecurity throughout 2016. In January 2017, however, food insecurity became a more serious issue in Kidal, and 19% of households were affected. In Timbuktu, few

5 Resident Enumerators for Continuous Monitoring

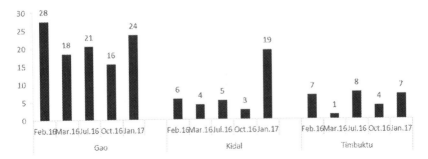

Fig. 2 Percentage of households living in a state of food insecurity (*Source* Authors' calculations based on data from the Permanent Monitoring System)

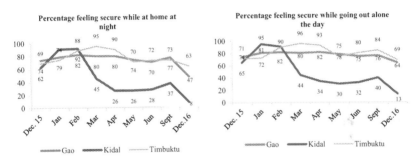

Fig. 3 Perceptions of security (*Source* Authors' calculations based on Mali Permanent Monitoring System)

were affected by food insecurity throughout the duration of the survey (Fig. 2).

Despite the Peace Agreement, the surveyed households' sense of security decreased considerably during 2016. Between January and December 2016, the percentage of the population who were comfortable at home at night decreased from 79 to 47% in Gao, from 91 to 8% in Kidal, and from 74 to 63% in Timbuktu. The pattern was the same for feelings of security during the day: Between January and December 2016, the percentage of population who felt secure going out alone during the day decreased from 81 to 64% in Gao, from 95 to 13% in Kidal, and from 72 to 69% in Timbuktu (Fig. 3).

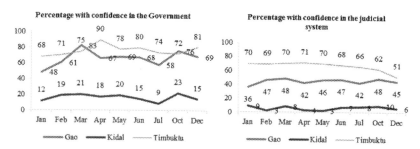

Fig. 4 Confidence in the government and the judicial system (*Source* Authors' calculations based on Mali Permanent Monitoring System)

Following the conflict, confidence in government was low, especially in Kidal, where less than 20% of the population was found to have confidence in the Malian government. Confidence levels barely changed throughout 2016. In Gao, the percentage of the population having confidence in the government never exceeded 70%. In Timbuktu, levels of confidence were generally higher, but they fluctuated quite considerably over time. Confidence levels were not much different in terms of the judicial system, with particularly low levels in Kidal, higher in Gao, and highest in Timbuktu, where over the course of 2016, confidence in the judicial system decreased from 70 to 51% from January to December (Fig. 4).

The patterns of confidence in the government and the judicial system carry over with respect to trust in people from other ethnic groups and foreigners. In Gao and Timbuktu, the percentage of the population with trust in people from other ethnic groups was relatively high compared to Kidal, but decreased over time, from 77% in Gao and 75% in Timbuktu in January 2016, to 71% in Gao and 56% in Timbuktu in December 2016. In Kidal, levels of trust were found to be significantly lower, at around 40% and falling during 2016. Trust in foreigners was virtually non-existent in Kidal, at less than 10%, but much higher (and rising) in Timbuktu and Gao. In contrast, the population across all three locations was found to have high levels of confidence in religious and traditional authorities. The quasi-totality of the population in

5 Resident Enumerators for Continuous Monitoring

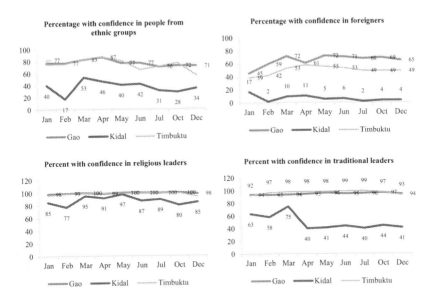

Fig. 5 Confidence in people (*Source* Authors' calculations based on Mali Permanent Monitoring System)

the three regions indicated that they had confidence in religious leaders, and the same pattern held true for traditional leaders in Gao and Timbuktu, but not in Kidal, where confidence in traditional leaders was found to be much lower, at 63% in January 2016, and declining over time (Fig. 5).

During 2016, the problems faced by healthcare centers remained largely unresolved. Some problems, such as the lack of medication and lack of staff, even increased between January and December 2016, although staff absenteeism declined considerably over the same period. Other problems, such as the lack of infrastructure, became less pressing, but generally speaking, very limited progress was made in restoring health services. The state of schools was similar. The lack of teachers decreased from 24 to 16% over 2016, as did the lack of school materials, declining from 24 to 20%. However, other issues became more pressing, including the lack of classrooms and the absence of school feeding (Fig. 6).

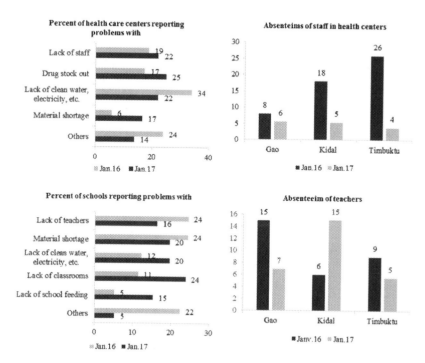

Fig. 6 Problems reported by health facilities and schools (*Source* Authors' calculations based on Mali Permanent Monitoring System)

4 Lessons Learned and Next Steps

Implementation of the PMS was surprisingly straightforward, not least because World Bank staff collaborated closely with a high-quality survey firm with experience in northern Mali. Two major challenges, the hiring of enumerators and ensuring data quality, have been discussed already. A third challenge proved to be financing. While the data produced were well-received and in demand, and even though each round of data collection was relatively inexpensive, after 15 months of continuous data collection, the team failed to identify the funding needed to continue the exercise.

Fortunately, an alternative was identified. While financing for generalized data collection proved hard to find, funding for third-party monitoring of project activities was available. The mix of terrorism and armed violence rendered field supervision by donor representatives impossible. At the same time donors desired to invest more in the north to support the peace process. Because donor representatives were not able to visit project sites in northern Mali, they started to rely on third-party monitors. Often, these are local NGOs that are also involved in reconstruction activities (raising concerns about conflicts of interest), or specialized outsider firms with a higher risk appetite, at a commensurate price. Irrespective of their nature, these third-party monitors collect information, for example on the progress of a construction project, which is a task familiar to the resident enumerators used for this project. The resident enumerators were thus retrained to act as third-party monitors. Relying on local enumerators for third-party monitoring is new, and the World Bank is testing this approach against an alternative of visits by experts from local NGOs. This is ongoing, but if the results of the continuous data collection in northern Mali offer any guidance, it seems likely that the local enumerators, equipped with tablets, cell phones, and motorbikes, will be able to provide quality data at a fraction of the cost that is usually paid for third-party monitoring.

Annex: Evolution of Security and Economic Indicators

78 A.-M. Taptué and J. Hoogeveen

Outcome	Indicators	Regions	Jan 2016	Feb	Mar	Apr	May	June	July	Oct	Jan 2017
Security is restored in the north	*Population's perception of their own security*										
	Percentage of villages/areas where police is permanently present	Gao	11	11	11	0	11	0	5	6	6
		Kidal	0	0	0	0	0	0	0	0	0
		Timbuktu	5	2	5	3	3	3	0	5	3
	Percentage of villages/areas where gendarmerie is permanently present	Gao	37	42	42	37	42	37	37	33	39
		Kidal	0	0	0	0	0	0	0	0	0
		Timbuktu	29	29	32	29	32	26	26	32	29
	Percentage of villages/areas where army is permanently present	Gao	48	53	53	42	47	42	42	45	45
		Kidal	0	0	0	0	0	0	0	0	0
		Timbuktu	38	25	34	37	30	32	33	32	21
	Percentage of population who feel secure while going out alone in the day	Gao	60	74	81	82	80	82	78	76	64
		Kidal	59	65	95	91	44	34	30	40	13
		Timbuktu	69	71	72	90	96	93	75	84	69
	Percentage of population who feel secure on road to the nearest village/quarter	Gao	40	32	30	29	34	45	46	63	44
		Kidal	44	23	21	63	74	77	77	29	3
		Timbuktu	61	61	42	32	40	60	59	59	43
	Percentage of households with school-age children who think that their children are safe on the way to school	Gao	77	68	72	76	66	63	53	75	39
		Kidal	30	60	74	29	12	0	6	20	4
		Timbuktu	76	70	82	92	94	95	88	92	72
The government's capacity to offer basic services have increased	*Capacity for service delivery*										
	Percentage of teachers present during the school day (relative to number on payroll)	Gao	85				86			91	92
		Kidal	94				81			100	85
		Timbuktu	91				91			91	95
	Percentage of health workers present (relative to the number on payroll)	Gao	92				74			90	94
		Kidal	82				95			90	95
		Timbuktu	74				87			93	96
	Average number of functional wells per village	Gao	3						8	8	7
		Kidal	2						3	3	5
		Timbuktu	2						3	3	9

5 Resident Enumerators for Continuous Monitoring

Outcome	Indicators	Regions	Period								
			Jan 2016	Feb	Mar	Apr	May	June	July	Oct	Jan 2017
	Use of basic services										
	Percentage of households with livestock who have used veterinary services in last month	Gao		3	6	9	2	3	3	15	25
		Kidal		4	0	4	5	0	2	3	2
		Timbuktu		10	13	15	13	12	13	16	27
	Percentage of eligible children (6–12 years) attending primary school	Gao	47			63	57	55	///	65	65
		Kidal	2			29	12	16	///	43	51
		Timbuktu	50			62	54	48	///	57	64
Peace has been reinforced through reconciliation and practical cooperation between local authorities and community	*Perception of confidence toward Peace Agreement and trust in government*										
	Percentage of citizens who express trust in government	Gao	83	86	89	67	69	68	58	89	69
		Kidal	30	25	28	18	20	15	9	24	15
		Timbuktu	87	93	94	90	78	80	74	87	81
	Percentage of citizens who express trust in police	Gao	38	57	74	57	61	62	58	68	61
		Kidal	10	17	16	13	18	8	9	22	17
		Timbuktu	40	50	48	53	57	59	57	59	60
	Percentage of citizens who express trust in the Army	Gao	48	64	84	67	69	67	59	71	64
		Kidal	12	17	16	13	18	9	9	22	17
		Timbuktu	65	74	79	82	76	81	74	70	80
	Percentage of citizens who express confidence in the peace process	Gao	82	90	95	96	94	92	48	75	57
		Kidal	68	40	49	45	22	27	12	37	14
		Timbuktu	77	60	95	95	90	82	65	74	74
	Percentage of citizens who express trust in local officials	Gao	81	91	91	89	89	91	86	89	85
		Kidal	39	29	57	42	39	46	45	38	46
		Timbuktu	80	87	84	84	81	76	81	79	72
	Percentage of citizens who express trust in the judicial system	Gao	36	47	49	42	46	47	42	49	45
		Kidal	9	3	9	4	3	7	8	10	6
		Timbuktu	70	69	71	71	70	68	66	63	51
	Tangible evidence of peace										
	Percentage of people who perceive positive improvements in their community since the start of the peace process	Gao	72	71	82	82	82	67	57	72	54
		Kidal	67	42	44	43	47	43	29	36	21
		Timbuktu	57	61	75	97	89	70	70	79	73

(continued)

(continued)

Outcome	Indicators	Regions	Period								
			Jan 2016	Feb	Mar	Apr	May	June	July	Oct	Jan 2017
Economic recovery and sustainable livelihoods are effective for all	*Perceptions of wellbeing*										
	Percentage of citizens who feel better off compared to last twelve months	Gao	41	23	17	13	12	22	21	20	13
		Kidal	40	28	29	27	16	18	7	34	19
		Timbuktu	51	33	38	48	54	47	43	39	38
	Percentage of citizens who expect to be better off in the next twelve months	Gao	68	53	53	38	35	38	37	37	38
		Kidal	44	6	14	11	17	11	15	31	14
		Timbuktu	75	66	76	72	67	68	57	55	62
	Average number of meals per day	Gao	3	2	3				4	2	3
		Kidal	3	3	4				5	3	2
		Timbuktu	3	3	5				3	3	3
	Percentage of households living in a state of food insecurity	Gao		28	18				21	16	6
		Kidal		6	4				5	3	19
		Timbuktu		7	1				8	4	15
	Percentage of those aged 15–60 who are employed for cash (wages) during the week preceding the survey	Gao	14								9
		Kidal	8								14
		Timbuktu	8								19
	Percentage of those aged 15–60 who report increased confidence in future employment opportunities	Gao	56	52	49	39	41	34	32	40	31
		Kidal	34	7	14	10	9	13	8	26	2
		Timbuktu	67	54	74	73	61	61	53	52	66
	Average price of kg of local husked rice	Gao	425	389	386	401	404	416	421	396	420
		Kidal	290	400							300
		Timbuktu	426	346	386	338	353	351	346	365	324
	Average price of kg of beef without bones	Gao	1935	2084	2070	1965	1838	2183	2039	2487	2541
		Kidal	2639	2225	1970	2250	2250	2143	2313	3125	2333
		Timbuktu	1822	1898	1904	1877	1893	1974	1896	1963	1918

Outcome	Indicators	Regions	Jan 2016	Feb	Mar	Apr	May	June	July	Oct	Jan 2017
Northern regions are integrated in the Malian economy	*Connectivity*										
	Percentage of localities with access to electricity	Gao	21					21	16	16	22
		Kidal	48					48	48	48	48
		Timbuktu	17					21	21	21	17
	Percentage of localities with cellphone network	Gao	84					89	89	84	79
		Kidal	48					48	48	42	48
		Timbuktu	84					84	84	84	84
	Percentage of localities connected to a radio antenna	Gao	58					63	53	61	50
		Kidal	28					9	0	19	9
		Timbuktu	41					53	41	53	57
	Percentage of localities connected to a TV antenna	Gao	53					58	63	56	39
		Kidal	28					0	0	0	10
		Timbuktu	41					49	41	41	45
	Percentage of localities having an Internet connection	Gao	16					21	21	17	22
		Kidal	28					28	38	28	38
		Timbuktu	21					28	25	33	29
	Average cost of transport to the center of the district (FCFA/kg)	Gao	26	41	34	30	46	45	35	55	43
		Kidal	100	56	51	48	51	56	62	65	52
		Timbuktu	78								
	Markets										
	Percentage of outlets where industrial fertilizers are available	Gao	50	6	9	9	10	11	15	13	100
		Kidal	0	0	0	0	0	0	0	0	0
		Timbuktu	37	25	24	13	20	16	26	16	13
	Percentage of big outlets where seeds of rice are available	Gao					19	38	48	18	
		Kidal					0	0	0	0	
		Timbuktu					3	5	7	6	
	Price of one liter of gasoline	Gao	719	619	616	558	860	656	888	475	594
		Kidal	736	822	857	878			856	804	800
		Timbuktu	725		747	711	697	713	810	616	633

Source Authors' calculations based on the Mali Permanent Monitoring System

References

INSTAT. (2012). 4ème Recensement Général de la Population et de l'Habitat du Mali (RGPH-2009). Analyse des résultats définitifs. Thème: scolarisation, instruction et alphabétisation au Mali.

World Bank. (2016). Sampling Frame for Mali Permanent Monitoring System. Mimeo.

The opinions expressed in this chapter are those of the author(s) and do not necessarily reflect the views of the International Bank for Reconstruction and Development/The World Bank, its Board of Directors, or the countries they represent.

Open Access This chapter is licensed under the terms of the Creative Commons Attribution 3.0 IGO license (https://creativecommons.org/licenses/by/3.0/igo/), which permits use, sharing, adaptation, distribution and reproduction in any medium or format, as long as you give appropriate credit to the International Bank for Reconstruction and Development/The World Bank, provide a link to the Creative Commons license and indicate if changes were made.

Any dispute related to the use of the works of the International Bank for Reconstruction and Development/The World Bank that cannot be settled amicably shall be submitted to arbitration pursuant to the UNCITRAL rules. The use of the International Bank for Reconstruction and Development/The World Bank's name for any purpose other than for attribution, and the use of the International Bank for Reconstruction and Development/The World Bank's logo, shall be subject to a separate written license agreement between the International Bank for Reconstruction and Development/The World Bank and the user and is not authorized as part of this CC-IGO license. Note that the link provided above includes additional terms and conditions of the license.

The images or other third party material in this chapter are included in the chapter's Creative Commons license, unless indicated otherwise in a credit line to the material. If material is not included in the chapter's Creative Commons license and your intended use is not permitted by statutory regulation or exceeds the permitted use, you will need to obtain permission directly from the copyright holder.

6

A Local Development Index for the CAR and Mali

Mohamed Coulibaly, Johannes Hoogeveen, Roy Katayama and Gervais Chamberlin Yama

1 The Data Demand and Challenge

The Central African Republic (CAR) has been affected by repeated cycles of violence and conflict. A landlocked country in Central Africa, with an area of about 620,000 square kilometers and an estimated population of around 4.9 million, the CAR is sparsely populated. Despite a wealth of natural resources such as uranium, crude oil, gold, diamonds, cobalt, lumber, wildlife, and hydropower, as well as significant quantities of arable land, the CAR is among the ten poorest countries in the

---

M. Coulibaly · J. Hoogeveen · R. Katayama (✉) · G. C. Yama
World Bank, Washington, DC, USA
e-mail: mcoulibaly2@worldbank.org

J. Hoogeveen
e-mail: jhoogeveen@worldbank.org

R. Katayama
e-mail: rkatayama@worldbank.org

G. C. Yama
e-mail: gyama@worldbank.org

© International Bank for Reconstruction and Development/The World Bank 2020
J. Hoogeveen and U. Pape (eds.), *Data Collection in Fragile States*,
https://doi.org/10.1007/978-3-030-25120-8_6

83

world. According to the Human Development Index, the country had the lowest level of human development in 2016, ranking last out of 188 countries.

The latest bout of insecurity started in late 2012 with a Séléka insurrection in the north of the country. This led to three years of violence, destruction of property, great human suffering, and left an estimated one-fifth of the population displaced. In May 2015, the Bangui Forum was organized to discuss the country's peace-building program, and to pave the way for elections. After another major outbreak of violence in September 2015, the country successfully held presidential and legislative elections in early 2016 and induced a lull in the conflict.

Despite the reduction in conflict, the country remained insecure even after the elections. More than a dozen armed militias remain active in the country today, controlling most of the country's territory. These armed groups are pursuing a wide spectrum of objectives. The Anti-Balaka, which arose from village-based self-defense groups, and the Union for Peace in the CAR (UPC), comprised mostly of Fulani cattle herders with the aim to protect transhumance corridors, have a strong focus on community protection. The Lord's Resistance Army (LRA), on the other hand, has no territorial or ethnic ties in the CAR, and uses the country as a safe haven and source of revenue through looting. The Popular Front for the Renaissance of the CAR (FPRC), by contrast, is active in the northern regions of the country and is closer to Chad. The United Nations peacekeeping mission (MINUSCA) operates among these armed groups. This mission, although unpopular, remains essential given the inoperative national defense and security forces, and the lack of state presence throughout the country.

To forge a national consensus on the country's needs and priorities for the first five years of the post-election period, in May 2016 the government of the CAR requested support from the European Union, the United Nations, and the World Bank Group to prepare a Recovery and Peacebuilding Assessment (RPBA). Those preparing the assessment were in urgent need of up-to-date information about the country, and requested data that could inform the planning of recovery activities and serve as a baseline for a monitoring system. The challenge was made greater by the fact that the new data and analytical results were needed by September 2016, leaving only three months to prepare and complete the data collection. Moreover, the rainy season was about to start and road infrastructure was in poor condition.

Household surveys take time to design and implement, and a typical welfare survey takes more than a year to prepare, field, and analyze. It was clear that a more adapted solution would be needed. To complicate matters further, the conflict had left the country's statistical system, which had been reasonably developed prior to 2012, in poor shape. Many staff of the national statistical institute (ICASEES) had left, its offices had been pillaged, and much of the country's statistical memory had been wiped out. The existing sampling frame was outdated and no longer reliable given that entire villages had vanished, and 20–25% of the population was displaced.

2 The Innovation

When considering the request for new statistical data, the team realized that given the precarious security situation, travel would need to be minimized. At the same time, disparities between Bangui and the country's periphery had been recognized as one of the drivers of the conflict, and thus collecting information nationwide was imperative. Donors also made it clear that poverty estimates should be updated as insecurity and massive internal displacement had made the existing poverty estimates less relevant for decision making; as such, new poverty maps were needed that could be used to target interventions. It was evident, however, that it would be impossible to field and analyze a consumption survey within the given timeframe. Moreover, in the absence of a reliable sampling frame, such a survey would not constitute value for money.

The 2008 poverty numbers showed that even before the crisis, poverty in the CAR was pervasive. Poverty levels were estimated at 66% of the population, based on the international poverty line of US$1.90 per day in 2011 purchasing-power parity terms. Since that time, the country's gross domestic product (GDP) per capita fell by one-third, and recent estimates suggest that the poverty rate surged to more than 76%

in 2015. When almost everybody is poor, further refining the number of people living in poverty is of limited value, and means-based targeting is not a key priority. Instead, identifying what had to be targeted where was of greater importance.

Instead of producing a poverty map, the team decided to map the state of the nation by making a rapid assessment of the public services that were available. Drawing from the experience of Mali's Indice de Pauvrete Communale (District Poverty Index, IPC), a district census was designed for the CAR, called the *Enquête Nationale sur les Monographies Communales* (ENMC).[1,2,3] Enumerator teams would interview representatives and other district leaders from each of the 179 districts, the lowest administrative unit, in the country, using a structured questionnaire.[4] Since it was clear that in many locations officials were absent, and to avoid nonresponse because of this, the enumerator manual did not prescribe which officials had to answer, only that a group of officials who were knowledgeable about the district capital (*chef-lieu*) and the district's largest villages had to be identified. While this strategy was successful in that information was eventually obtained for every district, very detailed information from specialists could not be collected and questions had to remain relatively general.

The district census collected information on conditions in all districts across the country, including on local infrastructure, access to information (radio, television, and phone network), health and education facilities, local governance, economic activities, conflict, security, and violence, and local perspectives on security and policy priorities. On the basis that respondents would have more accurate information on their immediate environment, the questionnaire focused primarily

---

[1]Observatoire du Développement Humain et Durable (ODHD), 2008. Profil de pauvreté des districts de Mali.

[2]While in this chapter the district census is emphasized, the ENMC also had a household survey component. More on this in Sect. 4.

[3]The instruments, data, and analysis of the *Enquête Nationale sur les Monographies Communales* (ENMC) can be downloaded from: http://bit.ly/2k7wFlq.

[4]The administrative divisions in the CAR are as follows: (1) prefecture, (2) sub-prefecture, and (3) district, referred to as *commune* locally. The 8 administrative subdivisions of the capital city, Bangui, were treated as districts in the ENMC. The district census was carried out at the district level in the CAR.

6 A Local Development Index for the CAR and Mali 87

on the situation in district capital to improve the reliability of the data collected. In addition, district officials were asked to list the ten largest localities in their district outside the district capital, and to indicate the presence of schools, health facilities, water points, electricity, mobile phone networks, refugees and displaced people, transport opportunities, and markets in each of these localities.

A district census had several advantages. Districts are the smallest administrative divisions in the CAR, and are thus at the forefront of service provision. No sampling was required, as all 179 districts were to be covered. The small number of observations needed for this census had other advantages. Logistical complexity was reduced, and only a small number of enumerators had to be trained and supervised. Data collection and data entry were fast, and analysis and reporting were straightforward and visually appealing, as much of the information collected could be presented in the form of maps. Last but not least, the overall cost was small,[5] facilitating regular repeats of the ENMC and thereby ensuring that the RPBA's request to create the basis for a monitoring system could be fulfilled.

To facilitate decision-making, information collected in the district census was reflected in the Local Development Index (LDI). This composite index combines a range of policy-relevant indicators into a single measure. It thus sheds light on district conditions in a straightforward and easy-to-understand way. Moreover, by covering the entire country, the LDI could serve as an alternative to a poverty map, with the added advantage that all the tracked indicators are actionable by decision makers. This allows decision makers to identify which districts are in greatest need of additional investment. Decision makers can also use LDI scores as a basis for budget allocations, with underprivileged districts receiving larger per capita allocations, thus facilitating the process of decentralization.

The indicators used to construct the LDI fell into three categories: local administration, infrastructure, and access to basic services.

---

[5] It cost US$180,000 to design, field, and analyze the ENMC. This covered the district census as well as the associated household survey.

Local administration was captured through indicators such as budget per capita (in local currency) allocated to the district, number of working staff at the local district government office, and presence of security forces (gendarmerie and police). The second pillar assessed the availability of basic infrastructure, including the presence of a mobile phone network and a banking system, and the transport cost per kilometer, as a proxy for mobility costs across the country. The third pillar measured the availability of basic services, such as public primary schools, health centers, sanitation systems, and clean water. These three pillars constitute the overall LDI. As there is no objective way for the different pillars to be weighted, and to keep the results tractable, each pillar was equally weighted in the final score: that is, the weight for each pillar was one-third. Within each pillar, some indicators were given a higher importance than others, and were therefore attributed different weights; however, each cluster of sub-activities, particularly health, education, and water, were assigned equal weights. Details of the weighting scheme are shown in Table 1.

3 Key Results

The district census brought the characteristics of different areas that are critical for development planning into a single database. It presented information about the agro-ecological zone, the main and secondary sources of income, the main crops grown, and whether there were any mining activities in each district. The census collected information about the presence of displaced people and whether NGOs were active in an area. It also collected information on infrastructure, such as roads and electricity, and service delivery, such as schools and health centers. Finally, the perceptions of local officials were collected on their development priorities and how the current situation differed compared to six months earlier.

The district census confirmed the dismal state of development in the CAR, and demonstrated the considerable variation that exists across districts. District administration offices were found to be understaffed and short of funding. In most districts, security personnel (police and gendarmes) were absent, and only 24 districts had 20 or more staff in the municipal office, with regular payment of municipal staff remaining a

Table 1 Local Development Index: components and weights

Sub-index	Weight	Indicators	Weight	
Local administration	1/3	2016 budget per capita in CFAF (census 2003 population data)	1/3	
		Number of staff in the local district government office (Mairie)	1/3	
		Security, gendarmerie or police	1/3	
Infrastructure	1/3	Cost of transport to Bangui (CFAF per km)	1/3	
		Mobile phone reception in the district center	1/3	
		Banking in the district center	1/3	
Basic services	1/3	Percentage of the ten largest localities in the district with functional primary public schools	1/3	
		District capital has a maternal health center	1/18	
		District capital has a hospital or a health center	1/18	1/3
		Percentage of the ten largest localities in the district with functional health centers	4/18	
		Presence of the national water company (SODECA) or local water distribution in the district capital	1/18	1/3
		Share of 10 largest localities in the district with clean water (public fountains, boreholes, or protected wells)	5/18	

Source Authors' visualization

problem. Moreover, 57 districts indicated not having received a budget allocation for 2016.

Access to infrastructure, including electricity, mobile phone coverage, banking services, and road networks, was found to be low. Only 15% of districts reported having electricity or some form of public lighting in the district capital, and only one of the 101 district capitals located in a rural area was found to be connected to the national electricity grid. Overall, only four in ten district capitals had at least one mobile phone provider in the district capital. Furthermore, only one in ten district

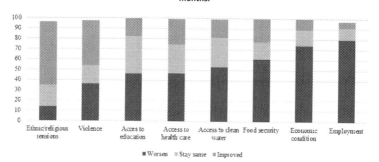

Fig. 1 Selected results from the district census (*Source* Authors' calculations based on the CAR District Census/ENMC)

6 A Local Development Index for the CAR and Mali

capitals had some form of banking system, either a bank or a local credit union. Half of the districts reported that roads to Bangui were not accessible throughout the year (Fig. 1).

Local administration: Funding and staffing in districts
There was low capacity in local governance, as districts lacked staff and funding. This was combined with the absence of gendarmerie and police forces in many districts.

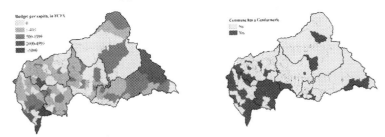

Local infrastructure: Mobile phone coverage, banking services, and roads
Essential infrastructure – e.g. mobile phone coverage, banking services, roads – was lacking in many districts.

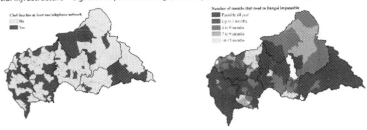

Access to basic services
Access to basic social services, such as primary schools and health centers, was found to be limited.

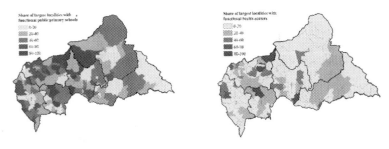

Fig. 2 Selected results on local administration, infrastructure, and access to services (*Source* Authors' calculations based on the CAR District Census/ENMC)

Access to basic social services such as public primary schools, health centers, and clean water was limited, especially outside district capitals. In the ten largest localities of each district, only 43% had a functional public primary school, 18% had a functional health center, and 43% had access to clean water sources. Access to clean water and sanitation systems were found to be limited even in the district capitals, where only 36% of the districts reported having clean water access points in their capitals (Fig. 2).

The LDI was constructed using the approach described in Table 1, shedding light on current conditions in a simple and straightforward way. The LDI score was low for most districts, indicating the need for substantial improvements across the country. Among the three pillars that form the LDI, local infrastructure varied more across districts, whereas access to basic services was relatively homogeneous. Compared to other districts in the country, those in Region 1, Region 2, and Region 7, which correspond to the capital and southwestern region of the country, were more likely to be in the top quintile of the LDI.

4 Implementation Challenges, Lessons Learned, and Next Steps

The ENMC demonstrated the feasibility of collecting nationwide information relevant to decision makers, both rapidly and in a cost-effective manner. The data informed project preparation and fed the RPBA monitoring system. Results have been widely disseminated, and representatives in each district have received posters showing how they perform relative to other districts in the country (Fig. 3). The district census will be repeated annually to track progress.

Because the main cost of most surveys is the transport cost for enumerators to physically reach the survey locations, the district census was supplemented by a light household survey at a marginal cost. The survey was considered 'light' in the sense that no detailed consumption data were collected. Sampling for the household survey took account of the fact that traveling throughout the country was still dangerous, and time was limited

6 A Local Development Index for the CAR and Mali

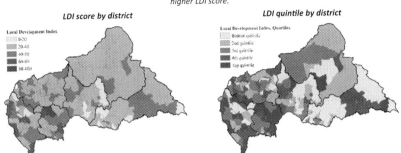

Fig. 3 Local Development Index across districts (*Source* Authors' calculations based on the CAR District Census/ENMC)

for data collection. Given these concerns, in addition to high transport costs, an unorthodox sampling design was selected in which ten households were interviewed in each district where five households were randomly selected from a randomly selected neighborhood of the chef-lieu, and five households were randomly selected from a randomly selected village located 20–40 kilometers from the chef-lieu. In each of the selected localities, a simple listing of households was completed, up to a maximum of 100 households, from which the five households were selected.

The survey was designed such that a team of two enumerators and a driver could collect all the information from one district within two days, allowing for speed of data collection, and reducing costs and exposure to risk. This strategy was successful. District officials from all 179 districts were interviewed, and in the end, households in only two districts could not be interviewed because the situation was too dangerous. Officials from these two districts were interviewed in neighboring locations. A total of 1767 households were interviewed.

The household survey served as a valuable complement to the district census. It allowed differences in perceptions and priorities for development between citizens and their representatives to be investigated, and the results show that these differences were minimal. Repeating both the household survey and the district census will aid in understanding whether improvements in service delivery as reported by district

Those with poor food consumption tend to be less wealthy and located in the two northern agro-ecological zone, which overlap with the Fertit, Yada, and Plateaux regions.

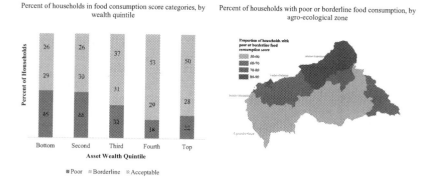

Fig. 4 Food consumption by wealth and agro-ecological zone (*Source* Authors' calculations based on the CAR District Census/ENMC)

representatives match improvements in outcomes, such as education and health, reported by households.

The household survey further allowed for the collection of information about wealth, displacement, the experience of shocks, the impact of the crisis, and food security. Using a concept borrowed from the World Food Programme, the Food Consumption Score (FCS) was calculated using information about the frequency with which nine different types of food had been consumed by the household in the past seven days.[6] The FCS was then used to explore which households found themselves in one of three categories: severely food insecure (poor), moderately food insecure (borderline), or food secure (acceptable) (Fig. 4).

In light of the situation, the ENMC was a success: the three-month deadline was met and a set of valuable data were generated, which informed and continues to inform decision makers. In an FCV context where state presence is limited and/or contested, the mere fact of collecting data nationwide contributed to a sense of equal treatment among districts and a feeling of belonging to one nation state.

---

[6]World Food Programme 2008. Food consumption analysis calculation and use of the food consumption score in food security analysis.

6 A Local Development Index for the CAR and Mali   95

This was important. The census was among the very few public initiatives which were successful in covering the entire territory and to which the Government could point as evidence of its commitment to all citizens across the nation's territory. A sample survey would not have had this intangible benefit.

With the benefit of hindsight, some aspects of the process could have been improved. More time could have been spent on developing the district census questionnaire, thus avoiding the need to change the contents of the questionnaire in its second wave (fielded in 2018) when the authorities were warming up to the idea of an LDI. On the other hand, once the initial LDI was constructed, it proved to be much easier to convince officials to substantively contribute to discussions about what it should entail.

While the team remains generally satisfied with the data collected by the household survey, it would have been advantageous if more households could have been interviewed in some areas. Bangui, the capital city, is comprised of eight administrative subdivisions (arrondissements), and the 78 households that were interviewed in Bangui were too few to support detailed reporting for the capital city. This also holds for some of the northeastern prefectures, which comprise very few districts and consequently, an insufficient number of observations were collected to support more disaggregated reporting. In addition, while the survey collected information from displaced people who were residing with extended families, camps for Internally Displaced People (IDP) were not covered by the survey.

Most importantly, the experience of ICASEES, the national statistical office, in fielding a survey in hard-to-reach and insecure areas was invaluable. Enumerators were given vests and their cars mounted with flags that demonstrated clearly that they worked for ICASEES, giving them some degree of protection from armed groups. Furthermore, enumerators assigned to at-risk areas were paid slightly more to motivate them to go and to avoid adverse selection in which the least experienced enumerators go to the most difficult areas. Trips were carefully planned, taking into account the type of infrastructure available and the appropriate means of transport. Where needed, motorbikes or boats were used.

Teams traveling into areas considered highly insecure were in regular contact with their team leader in Bangui. Although overall mobile phone coverage was limited to urban centers, it allowed teams to be followed closely as they moved from one location to the next. In some cases, teams borrowed radios from the UN or NGOs to contact supervisors. In addition, prior to the deployment, teams were trained to contact armed group leaders before entering areas controlled by them and to inform them about the data collection activity. Once in the area, the teams would pay a visit to these armed group leaders to seek their authorization in the form of a laisser-passer letter or stamp indicating their support for the activity. This allowed the teams to work in relative security, and the teams were escorted from the armed group in some cases in return for a small token of appreciation.[7]

Paper questionnaires were used, as tablets or smartphones were deemed too attractive to armed groups. UN flights were used to access hard-to-reach areas, where the teams often had to hire transport from local strongmen, giving them implicit protection. Teams received pocket money to be used at roadblocks to ensure safe passage. These measures turned out to be effective. Not only were all data collected in less than four weeks, but all teams returned to Bangui safe and unharmed.

Box 1   LDI in Mali allows for comparisons across time and space

For over a decade, Mali has conducted commune censuses which are similar to the Central African Republic (CAR) district census. While the CAR district census data are summarized in a Local Development Index (LDI), Mali's four commune censuses are used to compute the Indice de Pauvrete Communale (Commune Poverty Index, IPC) and poverty quintiles which are subsequently used in budget allocation formulas. The IPC is based on a principal component analysis (PCA) which is redone for every census, making the IPC noncomparable from one census to another.

---

[7]Clearly the presence of an escort by an armed group may have influenced responses to certain questions. Teams had been instructed beforehand to make sure that if they had an escort, armed group representatives would not be present at the interview.

6 A Local Development Index for the CAR and Mali 97

Taking advantage of the CAR experience, an LDI has been developed for Mali which allows comparisons of commune development across space and time. As in CAR, the LDI focuses on three aspects of development: local administration capacity, presence of infrastructure, and service delivery. These aspects have an equal weight of 33% and their sub-indicators are also equally weighted. The sub-indicators are common across all available commune censuses.

The LDI's definitions remain unchanged from one census to another and are comparable over time. The LDIs are positively correlated with the IPCs, and negatively with local poverty estimates. Because the LDI and CPI indices are ordinal, meaning that a lower value is associated with being poorer (IPC) or less developed (LDI), the (monotonic) relationship between them can be assessed using Spearman's correlation coefficient.[8] For the three first censuses for which both indices are available (2006, 2008, and 2013) the correlation coefficient lies above 0.65 with a p-value close to zero, suggesting a strong and statistically significant positive relationship. There is also a negative relationship between the LDI and the individual poverty rate (headcount ratio) of communes. The availability of a poverty map for Mali for 2009 made it possible to assess this relationship. Communes were grouped in two poverty categories depending on whether their poverty incidence was higher (first group) or lower (second group) than the national one. The LDI for poorer communes is significantly lower. Moreover, the LDI of the poor communes is lower than the national LDI average, which in turn is lower than the average LDI of the second group.

The new LDI is a useful tool for the analysis of development trends in Mali. For instance, looking at the regional LDI evolution between 2006 and 2017, Fig. 5 indicates that the communes in the region of Mopti and Segou had the highest increases in LDI (+76 and +61% respectively), while communes in the region of Kidal and Bamako had the lowest (+4 and +1% respectively). The big difference between Kidal and Bamako is that Bamako started at a very high base level, whereas Kidal started from a very low level. The LDIs show that before the crisis, the three northern regions were among the least developed in the country—lending support to grievances by the northern population about neglect by the central government. Broken down by livelihood zone, one notes that the progress in LDIs is strongly associated with crop production, and much less with nomadism and pastoralism.

The new index provides insight into the development dynamics of communes in the country. Figure 6, for instance shows the scatter plot of the

---

[8]A positive (negative) monotonic relationship between two variables is a relationship doing the following: as the value of one variable increases, the other variable value increases (decreases).

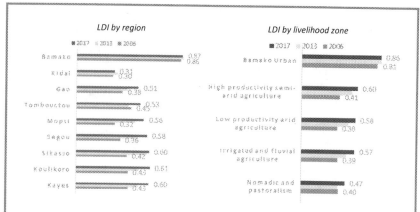

Fig. 5 Mali Local Development Indices, by region and livelihood zone (*Source* Authors' calculations based on the Mali Commune Censuses)

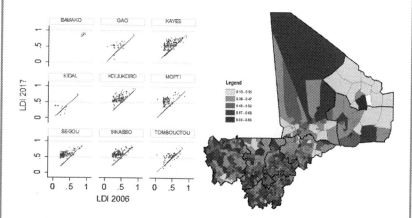

Fig. 6 LDI 2006 and 2017 by region and 2017 LDI map (*Source* Authors' calculations based on the Mali Commune Censuses)

LDI 2006 and LDI 2017 by region. It shows how the LDI for most communes improved (the dots lie above the 45-degree line), with the exception of Kidal and Tombouctou where a substantial fraction lies below the 45-degree line. The map demonstrates that almost all the worst performing communes can be found in the northern part of the country, and particularly in the North-East.

6 A Local Development Index for the CAR and Mali

The opinions expressed in this chapter are those of the author(s) and do not necessarily reflect the views of the International Bank for Reconstruction and Development/The World Bank, its Board of Directors, or the countries they represent.

Open Access This chapter is licensed under the terms of the Creative Commons Attribution 3.0 IGO license (https://creativecommons.org/licenses/by/3.0/igo/), which permits use, sharing, adaptation, distribution and reproduction in any medium or format, as long as you give appropriate credit to the International Bank for Reconstruction and Development/The World Bank, provide a link to the Creative Commons license and indicate if changes were made.

Any dispute related to the use of the works of the International Bank for Reconstruction and Development/The World Bank that cannot be settled amicably shall be submitted to arbitration pursuant to the UNCITRAL rules. The use of the International Bank for Reconstruction and Development/The World Bank's name for any purpose other than for attribution, and the use of the International Bank for Reconstruction and Development/The World Bank's logo, shall be subject to a separate written license agreement between the International Bank for Reconstruction and Development/The World Bank and the user and is not authorized as part of this CC-IGO license. Note that the link provided above includes additional terms and conditions of the license.

The images or other third party material in this chapter are included in the chapter's Creative Commons license, unless indicated otherwise in a credit line to the material. If material is not included in the chapter's Creative Commons license and your intended use is not permitted by statutory regulation or exceeds the permitted use, you will need to obtain permission directly from the copyright holder.

Part II
Methodological Innovations

7

Methods of Geo-Spatial Sampling

Stephanie Eckman and Kristen Himelein

1 Introduction

Technological advances in geospatial data have the potential to change how survey data are collected. Long hampered by high costs, limited capacity, and difficulties in supervision, sample selection is often done using second-best or nonprobability approaches. As geospatial technology has improved and become more widespread, costs have come down and the number of available tools have increased, making Geographic Information Systems (GIS)-based sampling approaches accessible to more users. This chapter presents experiences with GIS-based sampling from three different settings: (i) where no sampling frame is present because the census is outdated; (ii) sampling pastoralist communities; and (iii) rapid listing of enumeration areas to reduce exposure of field

---

S. Eckman
RTI International, Washington, DC, USA

K. Himelein (✉)
World Bank, Washington, DC, USA
e-mail: khimelein@worldbank.org

© International Bank for Reconstruction and Development/The World Bank 2020
J. Hoogeveen and U. Pape (eds.), *Data Collection in Fragile States*,
https://doi.org/10.1007/978-3-030-25120-8_7

104 S. Eckman and K. Himelein

teams. The case studies focus on extreme situations, particularly those in conflict-prone areas, as innovation often takes place when few other options are available. The applications discussed here, however, are applicable to many less extreme situations.

2 Data Challenge and Innovation #1: Creating a Sampling Frame in the Absence of a Census

For many studies, no sampling frame of the target population is available. The most common approach to addressing this problem for large-scale household surveys in the developing world is to use a stratified two-stage design. In the first stage, census enumeration areas are selected as the Primary Sampling Unit (PSU), using probability proportional to estimated size. In the second stage, a household listing operation is conducted in the selected PSUs, and households are selected using simple random sampling.[1,2] With this approach, even outdated census data can be used to select PSUs, as long as a high-quality listing operation is done in the selected PSUs to create a sampling for the second stage selection of households. Using out-of-date census data as a measure of size in PSU selection will result in estimates that are inefficient but still unbiased. However, some countries do not have census records at all because of accessibility issues, war, or natural disasters. In these situations, newly available high-resolution satellite data can be used to generate estimated population densities and to demarcate PSU boundaries. The two examples discussed here are from surveys conducted in rural Somalia and Kinshasa, Democratic Republic of the Congo (DRC).

---

[1]For the purposes of this chapter, the word "dwelling" is used to denote a physical structure inhabited by one or more households, while a "household" is a group of individuals that function as an economic unit. All methods that select dwellings which have the possibility of containing multiple households have selection protocols to randomly select an individual household for interview.

[2]Grosh and Muñoz (1996).

In Somalia, the last population census, carried out in 1975, measured the population at 3.9 million. Current estimates for the country indicate a population of more than 14 million. For the DRC, similarly, the last census was carried out in 1984, at which time the population was around 29 million. Current population estimates are now over 77 million. As noted above, it would still be possible to use the outdated census for estimated population totals if there was an expectation of approximately constant growth across regions. Both Somalia and DRC, however, have experienced significant civil strife, including large-scale displacements of the population.

Some countries, notably Haiti following the 2010 earthquake, have used "quick counts" to collect information about where the population lives and to estimate its size. In a quick count, enumeration areas are randomly sampled and listed, then the results are used to build a model to update census counts in the remaining areas.[3] However, in Haiti, the most recent census was only seven years old at the time of the earthquake, and the damage and population movements were relatively concentrated. The more time that has elapsed since the last census, the more difficult it is to develop an accurate model of the current population based on quick counts. Moreover, the DRC has a land area nearly 85 times the size of Haiti, which makes using a quick count methodology impractical from the perspective of both cost and implementation time. In Somalia, in addition to ongoing insecurity in certain areas, the enumeration area estimates from the 1975 census were never published, and the full results are thought to be lost. Therefore, alternative approaches to selecting a household sample were needed in both Somalia and the DRC.

2.1   Innovations

Three approaches were implemented across the two surveys. For the Somali High Frequency Survey (SHFS), rural areas posed a challenge for the creation of a sampling frame. Rural areas were defined as non-urban permanently settled areas but excluding Internally Displaced

---

[3]IHSI et al. (2012).

Persons (IDP) settlements.[4] To create a frame for the first selection stage, a gridded population approach was developed in collaboration with Flowminder.[5] Rural areas that were secure enough for data collection were divided into 100 by 100 meter grid cells. For each cell, WorldPop data provided an estimated population size.

Neighboring cells were then combined to form PSUs, using a quadtree algorithm, which combines cells to meet specified criteria, in this case, area and population size.[6] The maximum area was set at 3 by 3 kilometers, and the maximum population was limited to 3500 to keep enumeration areas manageable for field teams. The left panel of Fig. 1 shows the PSUs created by the above steps, with the color indicating the estimated population in each one.[7] Next, a sample of PSUs selected using probability proportional to estimated size. The selected PSUs were then further subdivided into segments. If the selected PSU contained 12 or fewer dwellings based on satellite imagery, only one segment was defined. For those PSUs containing between 13 and 150 dwellings, 12 segments were defined, with additional segments being defined for PSUs with more than 150 dwellings.

A major disadvantage of the grid approach described above is that the boundaries of the resulting PSUs do not follow natural boundaries such as roads, valleys, and rivers. The cells' artificial boundaries complicate field implementation. Aware of this constraint, the team initially pursued an alternative methodology in which the WorldPop distribution was used to randomly select points to serve as "seeds" for PSUs, which were then grown until they reached an estimated population of around 150 dwellings but without crossing natural boundaries.[8] Unfortunately,

---

[4]In urban areas, boundaries and population estimates were available from the United Nations Population Fund's Population Estimation Survey. Boundaries of IDP settlements were provided by United Nations High Commission for Refugee's Shelter Cluster.

[5]Closely following the methodology by Muñoz and Langeraar (2013).

[6]See Samet (1984), for a description of the methodology, and Minasny et al. (2007), for an application of the methodology to sample design.

[7]The map shows both urban and rural areas. Urban areas were not subject to the same population or land area limits.

[8]Thomson et al. (2017).

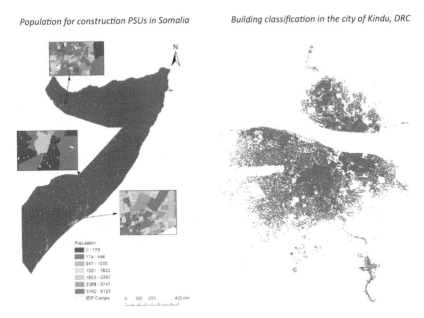

Fig. 1 Building classifications (Color figure online) (*Source* Authors' calculation)

two major drawbacks became immediately apparent: the development of algorithms to detect natural boundaries was expensive and time-consuming, and selection probabilities were not straightforward to calculate because of boundary effects (seeds near boundaries could grow in fewer directions than others). The team therefore reverted to a gridded approach but manually adjusted segments to follow natural boundaries to mitigate potential implementation issues.

In the DRC, two methods were used. In the districts of Kisenso, Kimbanseke, and Mont Ngafula in Kinshasa, and the sites of Kindu, Tchonka, and Basankusu, a one-stage sample of dwellings was selected based on counts of dwellings made from satellite images. In partnership with the firm Satplan Alpha, the project used recent satellite images to count and geo-locate all dwelling units. This work was done manually. Team members classified each building in the satellite images as low-density residential, high-density residential, or non-residential, using their local knowledge of the typical characteristics of dwelling units in the DRC. These typical characteristics were locally specific, varying between cities and between dense inner-city districts, peri-urban zones, and semirural areas on the outskirts. The main characteristics used to classify structures were architecture, building size and features, roof segmentation, roof design intricacy and height, building orientation, site boundary features, proximity to major streets, street activity, and traffic. The right panel in Fig. 1 shows the final map for Kindu, DRC, with each building classified as low-density residential (blue), high-density residential (yellow), and non-residential (red). When the counting, geo-locating, and classification were complete, each dwelling was assigned a random number, and a sample was selected through a one-stage random draw. If the classification was correct, this approach resulted in an equal-probability simple random sample of dwellings.

In the districts of N'djili and Makala in Kinshasa, a two-stage random sample was used.[9] PSU boundaries were first defined using administrative and physical boundaries such as rivers, highways, and secondary and residential roads that would be easily identifiable by interviewers on the ground. The delineation process used an automated iterative approach where PSUs were created and then split or merged based on target population size. The left panel of Fig. 2 shows a map indicating the manually created PSUs.

The next step was to estimate the population within each of these PSUs from high-resolution satellite data. First, a Random Forest Regression model was used to estimate population density based on contextual image information (image metrics that incorporate various aspects of surrounding information, rather than single-pixel signature).[10] The model was trained using a sub-sample of building locations.[11] The area and average building density for each PSU was then integrated with land use and land-cover data to adjust the area by the percentage covered with vegetation and then to produce a building count.[12]

---

[9]For further information, see Hirn and Rodella (2017).

[10]Implemented using MapPy, a Python library for remote sensing developed by Jordan Graesser.

[11]Graesser et al. (2012) provides a more detailed description of contextual image information in image processing.

[12]Building counts derived in this way produce comparable results to manual rooftop counts.

PSUs defined by natural and administrative boundaries   *Algorithmic estimate of population density*

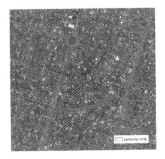 

Fig. 2 Boundaries and population densities (*Source* Authors' calculation)

PSUs were selected with probability proportional to this estimated size. A full listing operation was then conducted in the selected PSUs prior to the second stage selection of households. This approach leads to estimates with larger variances, and therefore less precise estimates, than the single-stage approach because the resulting sample is clustered.[13]

2.2 Key Results and Implementation Challenges

Each of the methods described above produced a sampling frame from which a representative sample was selected. There were, however, substantial challenges in Somalia. For the SHFS, 407 PSUs were selected for the survey (320 urban and 87 rural), and 366 PSUs were selected as replacements (251 urban and 115 rural). After selection, the PSUs were overlaid with satellite imagery from Google Earth and Bing to verify the presence of dwellings. Following that process, 53% of rural PSUs and 2% of urban PSUs were discarded and replaced due to having no visible

---

[13]Eckman, S., and B. West (2016), "Analysis of Data from Stratified and Clustered Surveys," in Wolf, C., Joye, D., Smith, T., and Fu, Y. (Eds.), *Handbook of Survey Methodology*. Thousand Oaks: Sage, 477–487.

population. In some cases, it was necessary to replace a PSU multiple times before one with visible dwellings was identified.

The approach used in the DRC generated more reliable results. Both the single-stage and multi-stage methods yielded results close to what the interview teams found during the listing exercise. The single-stage approach, which manually located dwellings based on satellite imagery and then drew a one-stage random sample, was applied in three large districts of Kinshasa. Locating individual dwellings on satellite imagery remains a manual task that is both relatively time-consuming and cannot be entirely standardized. While guidelines can be set for identifying dwellings, in practice, judgment calls are often required to (for example) distinguish businesses or separate conjoined structures into multiple dwellings. When selected structures turned out to be businesses, empty or destroyed houses, or other non-dwelling structures, the misidentified structures were replaced by a randomly selected replacement dwelling. If such misidentification is not excessive and does not systematically vary across the sampled area, the sample can be assumed to remain unbiased. However, misidentification can increase costs and needs to be monitored closely. Systematic variation in the misidentification of households across the sampled area may bias the sample (for example, underrepresenting areas with many high-rise buildings if the true number of dwellings within high-rises is systematically under-identified in a rooftop count). From a practical point of view, interviewers also sometimes struggled to find the selected households in dense areas, because no addresses were available, only a rooftop view with a GPS point. This drawback can be mitigated, however, by equipping interviewers with GPS-capable phones and clear walking maps that point out local landmarks and house characteristics to help with identification.

The second approach used in the DRC, which first defined PSUs and then algorithmically estimated population numbers to allow for an unbiased two-stage selection, posed different challenges. First, refining the algorithm that estimates population density is technically more complex than a simple visual count of dwellings based on satellite imagery. Once in place, however, it can quickly create automated population estimates for large areas. A second challenge is the loss of statistical efficiency inherent in the two-stage approach. Third, interviewers carrying out the listing within selected PSUs sometimes struggled to follow PSU boundaries and to distinguish which buildings were within or outside a given PSU. To minimize such problems, it is critical to prepare clear walking maps for interviewers and guidelines on how to deal with overlapping properties.

In the 28 PSUs in the Makala municipality in the Funa district of Kinshasa, both manual counting of residential buildings (the first method) and the modeling approach (the second method) were used, permitting a comparison between the two methods and the actual number of households identified in the field listing. Compared to an actual total of 9322 households recorded by the listing, the manual approach identified 7489 dwellings, while the modeling approach generated 10,667 dwellings in the same area. The correlations between the estimated and the actual values at the PSU level were 88.7 and 93.1% for the manual approach and the modeled approach, respectively. This important result indicates that the algorithm outperformed manual counting, at least for this application. See Fig. 3 for

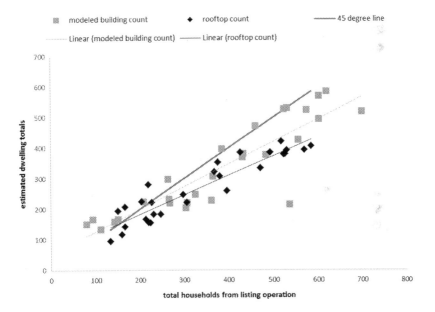

Fig. 3 Listing totals, modeled estimates, and rooftop counts for Makala (*Source* Authors' calculations)

a comparison of the dwelling counts estimated by the two methods with the household totals generated in the listing operation, for the 28 PSUs in Makala.

3 Data Challenge and Innovation #2: Sampling Pastoralist Communities

Livestock ownership serves a diverse set of functions in the developing world, from food source to savings and use as an investment vehicle. The pastoralist sector, however, has recently come under increasing pressures from several sources, including an increased demand for meat and dairy products from expanding middle classes, climate change, and the loss of traditional pasture land to development. Those who are among the most vulnerable to these pressures are nomadic and semi-nomadic pastoralist populations, but the transitory nature of their living situation also hampers the collection of high-quality representative data on which to base analyses.

Because many pastoralists lack a permanent dwelling, they are excluded by a traditional two-stage sampling approach. In July and August 2012, the World Bank undertook a survey in the Afar region of Ethiopia to test a novel approach to sampling the general population, including pastoralists.

The Afar region was selected for the pilot project for several reasons. First, the World Bank has an ongoing relationship with the Ethiopia Central Statistics Agency (CSA), including supporting the implementation of the Ethiopia Rural Socioeconomic Survey, which includes a module on pastoralist issues. The CSA also has a high-quality existing GIS infrastructure and a relatively high level of training compared to other potential study areas. Third, the Afar region offered geographic advantages over other pastoralist areas. It covers a land area of approximately 72,000 square kilometers in the north-east of Ethiopia and is relatively isolated. Well-guarded national boundaries, geographic features, and traditional ethnic hostilities limit the migration of the Afar people outside the boundaries of the region.

3.1 Innovations

The approach used to ensure pastoralist populations were included was the Random Geographic Cluster Sampling (RGCS) method. In an RGCS design, points (latitude and longitude) are randomly selected, and then a circular cluster of a given radius is created around the central point. All eligible respondents found within this cluster are selected for the survey. The main advantage of this design is that it captures everyone who is inside the selected circle at the time of the survey, including those who do not have a permanent dwelling or who are temporarily away from their usual dwelling. Properly implemented, this design eliminates the underrepresentation of mobile populations. Similar methods are commonly used in both developed and developing world contexts to measure agricultural production and livestock.[14]

To increase efficiency and lower fieldwork costs, the Afar region was divided into five strata, defined by the expected likelihood of finding herders and livestock. Spatial datasets describing land cover, land use, and other geographical features were used as input to delineate five discrete, mutually exclusive strata. The first stratum consisted of land in or near towns; the second stratum consisted of land under permanent agriculture; the third stratum was considered to be the most likely to contain livestock, and consisted of land within two kilometers of a major water source, including the Awash River and its permanent tributaries, and which also met the criteria for pasture based on a vegetation index; the fourth stratum consisted of land between two and ten kilometers from a major water source which also met criteria of pasture land; and the fifth stratum consisted of the remainder of the land area, which was considered to have the lowest probability of finding livestock (Fig. 4).

A total of 125 points were selected from these five strata for the survey. The number of selected points was higher in the strata with the highest expected concentrations of potentially nomadic households and livestock (Stratum 3) and lower in areas of lower expected

---

[14]For a more complete list of previous applications, see Himelein et al. (2014).

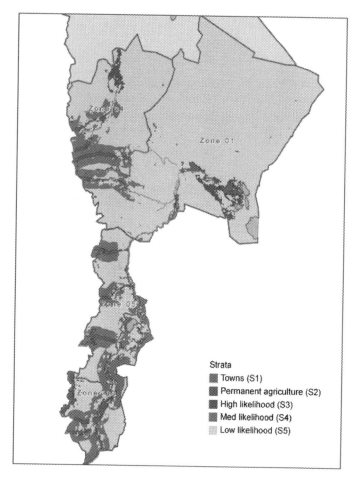

Fig. 4 Stratification map

density (Stratum 5). The radii for the circles also varied across the strata. In areas with higher expected densities, smaller circles were used to keep the workload manageable. In areas where few or no livestock were expected, the circle radius was expanded to the largest feasible dimensions to maximize the probability of finding animals. Table 1 lists the definition, sample size, and radius used in each of the five strata.

7 Methods of Geo-Spatial Sampling   115

Table 1 Stratification of the Afar region

Stratum	Description & excepted likelihood of finding individual/livestock	Radius (km)	Points selected	Total area (km²)	Percentage of total landscape
1	High likelihood: towns	0.1	10	33	<1
2	Almost no possibility: settled agricultural areas/commercial farms	0.5	15	930	2
3	High likelihood: within 2 km of major river or swamps	1	60	3538	6
4	Medium likelihood: within 10 km of major river or swamps	2	30	6921	12
5	Low likelihood: all land not in another stratum	5	10	45,152	80
	Total		125	56,574	100

After the selection of the PSUs, teams were given maps and hand-held GPS devices to conduct the surveys. Upon arriving at the center of a circle, the team canvased the circle and interviewed all households within its boundaries. The GPS device showed the selected circle, and alerted interviewers when they crossed into or out of area.

3.2 Key Results

The pilot project of the RGCS technique to collect livestock data in the Afar region of Ethiopia demonstrated that the implementation of such a design is feasible. Of the 125 points selected, 102 were visited. Of those visited, 59 circles (58%) contained at least one livestock animal. In total, the interviewers collected information from 793 households that owned livestock, although nine of these households were shown by their GPS coordinates to be outside of the circle boundaries and were there-fore excluded from the analysis, leaving a total sample size of 784. The number of interviewed livestock-owning households per circle ranged from one to 65, with a mean of approximately 15. In total, 3698 indi-viduals living in households owning livestock were identified as part of the survey. Of these, 127 reported having no permanent dwelling, which is a weighted estimate of 4701, or 2% of the livestock-holding population in the study area. All but five of the individuals without a permanent dwelling lived in households in which all members were completely nomadic. The inclusion of households without permanent addresses in the survey was a primary objective of the original research agenda because this group is traditionally underrepresented in dwelling-based surveys.

Overall, the project showed that sufficient GIS information is available, often in the public domain, to create strata for the probability of finding livestock, and to select points within those strata. With maps and relatively inexpensive GPS devices, interviewing teams can navigate the selected circles and identify eligible respondents within these clusters. The identified respondents can then be interviewed regarding their household's socioeconomic conditions and livestock holdings, creating the linkages necessary to understand the socioeconomic situation of these populations. In addition, using standard statistical methods, it is possible, although challenging, to calculate weights that take into account the varying probabilities of selection and that sufficiently address overlap probabilities. Moreover, information generated as part of the GPS field implementation can be used to account for underrepresentation, as discussed below. Finally, the methodology did what it was designed to do: Capture households without permanent dwellings that would have been excluded from a traditional dwelling-based sample design. The identification and interviewing of these households proved to be a major benefit to the RGCS, compared to the traditional household-based approach to survey sampling.

3.3 Implementation Challenges

Because the study area encompasses some of the harshest terrains in the region, and the methodology was novel for both the research and implementation teams, several unexpected difficulties were encountered. First, seasonal rains started earlier than expected, which created access problems such as the flooding of roads and land bordering the rivers.

7 Methods of Geo-Spatial Sampling 117

The access issues necessitated longer walks for interviewers, including one incident where a team had to walk 15 kilometers to reach the selected site. Other physical obstacles such as national park boundaries, active volcanoes, and militarized areas further restricted access to some locations. Third, ongoing strained relations between local communities and the national government led to a few isolated security incidents, including minor assaults against drivers and fieldworkers, and the (brief) kidnapping of the survey coordinator.

Beyond the implementation challenges, two other substantial issues arose as part of the data analysis process. The first was related to the calculation of the weights, which was much more complicated than originally anticipated.[15] The second challenge related to interviewers not canvasing the entire circle, and therefore missing potentially eligible respondents. The Viewshed analysis in Fig. 5 shows the path covered by the interviewers (the white lines), the portions of the circle they could have observed during their work (green and brown terrain map), and the black squares are the areas the interviewers could not have observed based on their path of travel. Several explanations for interviewers' failure to cover the entirety of the assigned circles are possible. The weather was extremely hot during this period. Flooding made access more difficult by requiring interviewers to take long detours on foot or ford swollen rivers. The survey took place during Ramadan, which limited the availability of local guides to assist the teams. Alternatively, however, it is feasible that the areas not observed were missed because they could not possibly contain any livestock, for example, because of the presence of flood water or vegetation too thick to traverse. Thus, the areas might be missing at random or not at random, and these two possibilities require different treatment in the analysis. Because it was impossible to distinguish the cause for the missed areas, two sets of statistics were reported for this study. This issue should be investigated closely for future implementations using this method.

---

[15]A full discussion of the correct procedure to derive probability weights is included in Himelein et al. (2014).

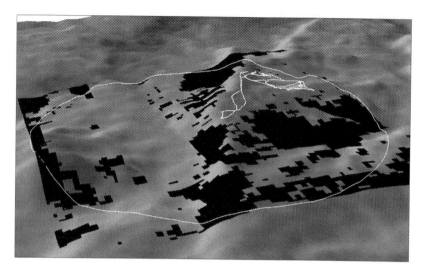

Fig. 5 Viewshed analysis (Color figure online)

4 Data Challenge and Innovation #3: Rapid Listing of Enumeration Areas[16]

The main challenge encountered in the Mogadishu High Frequency Survey (MHFS) Pilot[17] was related to security issues, which made traditional listings of households within PSUs impossible. The MHFS was conducted between October and December 2014 by the World Bank and Altai Consulting. In this case, the PSUs were selected from existing census enumeration area maps using probability proportional to estimated size according to the United Nations Population Fund's Population Estimation Survey. In the second stage of the survey, however, carrying out a full listing was deemed unsafe. Listing households in a PSU would require the team to spend an entire day in one

---

[16]See Himelein et al. (2017) for more complete discussion of the context and analytical approaches, as well as for the complete set of results.

[17]The MHFSA is a different survey than the Somalia High Frequency Survey discussed in Sect. 1.1.

7 Methods of Geo-Spatial Sampling   119

neighborhood, moving in a predictable pattern to reach all dwellings. The team's prolonged presence on the ground would increase their exposure to robbery, kidnapping, and assault, and increase the likelihood that local militias would object to their presence. A random walk procedure was initially proposed as a replacement, but this method has been shown in the literature to have a high likelihood of generating biased results, even if implemented under perfect conditions.[18]

The team considered four alternatives to a random walk. The first was to use satellite mapping to count rooftops. This methodology is shown in the right panel of Fig. 1 and discussed as the one-stage method used in the DRC survey above. The second alternative was segmentation, also shown in the left panel of Fig. 2: the creation of clusters with discernible boundaries on the ground. The third, grids, is discussed above. The fourth alternative was a novel proposal based on a random point selection methodology, but one that considers differing probabilities of selection generated by the spatial distribution of dwellings within a PSU.

Because Mogadishu at the time was deemed too dangerous to conduct pilots of the different methodologies, a comparison between the methods was made using a simulation study. The study simulated repeated sampling via the five methods described above in three purposefully chosen PSUs which varied in size, population, and socioeconomic status. Figure 6 illustrates the size and location of the selected PSUs.

To simulate the sensitivity of each method to different degrees of clustering, three methods were used to assign consumption values to the dwellings. In the first approach, values were randomly drawn from the distribution and assigned to dwellings in each PSU, resulting in no clustering in the consumption values. In the second and third approaches, the same values were reassigned within each PSU to create a moderate and a high degree of spatial clustering. After assignment, each dwelling in each of the three PSUs had three assigned consumption values.

---

[18]Bauer (2014, 2016).

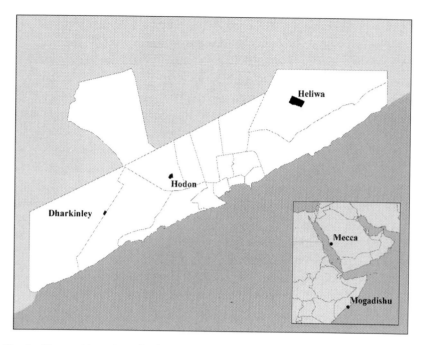

Fig. 6 Size and location of selected PSUs

4.1 Innovations

Several surveys have used random point selection methodologies to select households. In these methods, a random starting point is selected, and the interviewer is instructed either to interview the nearest dwelling or to proceed in a set direction until a dwelling is reached. The main drawback of these approaches is that the weights are difficult to calculate. Many researchers assume that the resulting sample is equal probability,[19] but that is not the case. A dwelling in a large open space has a higher probability of selection than one located in a densely-populated area: More points lead to the selection of the isolated dwelling.

The innovation proposed as part of the Mogadishu survey was to calculate the size of the "shadow" of the dwelling and use this information

---

[19]For further discussion, see Grais et al. (2007) and Kondo et al. (2014).

7 Methods of Geo-Spatial Sampling   121

to estimate the probability of selection. Interviewers were instructed to travel to each preselected point within the PSU, walk in the direction of the Qibla (the direction of Mecca), and to interview the first dwelling they reached. They repeated this approach until a sample of 10 dwellings had been achieved. The Qibla was used in Mogadishu because many interviewers have an app on their cell phones that indicates this direction, but any verifiable direction (north, south, etc.) would work similarly well. The probability of selection of each dwelling is proportional to the size of its "shadow": the set of all possible points that would lead to the selection of that dwelling. Figure 7 provides a visual representation of a dwelling's shadow in the Qibla method. Other random point selection methods lead to differently shaped shadows, but the principle is the same.

A major potential drawback of the Qibla and other related methods is the difficulty of measuring the area of the shadows. If high-quality, up-to-date satellite maps exist, then it is possible to use these images to calculate the shadow of a dwelling. The size, however, would be distorted if new structures had been built or demolished since the image was taken. Calculating the area of the shadow in the field could possibly be done by

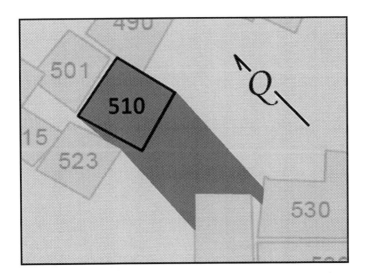

Fig. 7 Example of the Qibla method

asking the interviewer to walk the perimeter of the shadow with the GPS, but this would require substantial training, and may lead to measurement error. It would also increase the time spent in the field, which was not an option in an insecure context like Mogadishu. Two alternatives were therefore used to develop a proxy for the size of the shadow: The distance to the next structure in the opposite direction to the Qibla multiplied by the actual width of the dwelling, and the measured distance to the next structure multiplied by a categorical shadow width variable (small/medium/large) as defined by the interviewer. In addition, the simulation tested an approach which ignored the probabilities of selection and assumed the Qibla method led to an equal probability sample of dwellings.

4.2 Key Results

Figure 8 presents the results from the simulations of the five sampling methods. In this figure, the three PSUs are combined, but the sampling methods are shown separately. The points are the means of the sampling distributions

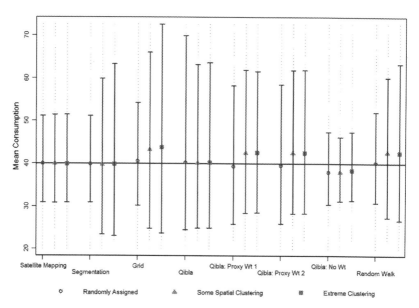

Fig. 8 Mean and confidence intervals (by method)

and the bars indicate the 5th and 95th percentiles. For each sampling method, there are three results shown: one for random assignment of consumption to dwellings (that is, no clustering); one for some clustering; and one for high clustering. The horizontal line at 40 is the true population mean consumption level. The results allow us to compare the methods' performance in terms of bias and variance and robustness to clustering.

The satellite method is unbiased: The mean over all the samples is the same as the true mean. This method is also unaffected by clustering in the consumption variable. These results were expected, because this method was assumed to be equivalent to the gold standard method of a full in-field listing; that is, the images were assumed to be up-to-date. Segmentation also showed consistently unbiased results, but higher variances for higher degrees of clustering in the underlying distribution, which is consistent with sampling theory on clustering.[20] The grid method, despite being conceptually similar to segmentation, overestimated the means with a bias up to 10% for the clustered distributions. The bias is related to grid squares that did not have enough dwellings to meet the sample size.[21] As expected, the Qibla method with the correct weights yielded unbiased results, but with wide confidence intervals, although these were partially driven by a few outliers. The values of the 5th and 95th percentiles of the distribution for this method are similar to those in the segmentation method when clustering is applied. The two methods of estimating the measure of size for the Qibla method showed a small amount of bias, ranging between 1.5% and 6.5%, depending on the degree of clustering. The final Qibla alternative, the unweighted version, consistently underestimated the true mean. The random walk approach, as noted above, is not theoretically unbiased, and this is reflected in the simulation results.

The main lesson learned from the simulation experiments is that full listing and satellite mapping generate the most consistently precise and unbiased results. It is also possible to generate unbiased results using a random point selection method—in this case, the Qibla method—but this approach

---

[20]See Eckman and West (2016).
[21]For a full discussion, see Himelein et al. (2017).

requires the accurate calculation of the area of the shadow to generate correct probability weights. If such complete data were available to researchers, satellite mapping would likely be a better choice. The Qibla method and segmentation are both unbiased and offer roughly similar precision, and therefore any choice between them would be based mainly on ease of implementation and the amount of information available. The other methods considered, including the proxy weights for the Qibla method and gridding, introduce some bias, but may be acceptable if other alternatives are not feasible. The two unweighted methods, the unweighted Qibla method and the random walk method, demonstrate the most bias, and should therefore be avoided.

5 Implementation Challenges

Because this study involved simulations and no fieldwork, fewer challenges were encountered. As discussed above, the main implementation challenges encountered for the Qibla method were related to the calculation of the shadow area, and by extension, the sample weights. In addition, some issues were encountered when the pre-selected point did not lead to the selection of a dwelling within the boundaries of the PSU. This issue was more pronounced in PSUs that had more open space, particularly on the perimeter of the city. In an actual survey environment, fieldwork protocols and training would be necessary to ensure consistency in addressing these situations.

6 Lessons Learned and Next Steps

It is clear from the accelerating pace of the application of GIS-based technology to sample design that the field will continue to expand in the coming years, driven by less expensive and higher resolution imagery and the development of better algorithms. Despite the excitement that these advances generate, however, researchers and practitioners must not lose sight of the importance of calculating accurate probabilities of selection to generate unbiased estimates. As shown by the RGCS design and the Qibla method, these calculations can be

7 Methods of Geo-Spatial Sampling 125

challenging, and there is a need for two complementary research areas in GIS-based sampling. The first area where research is needed is in improved population estimates where there is no census (or equivalent) frame. The work described above in the DRC and rural Somalia was a step in this direction. Flowminder, Facebook, WorldPop, and other groups have released population estimates.[22] The second research area is in relation to new methods of household selection when listing is not possible. The experiments in Afar and Mogadishu offer two alternatives. Unfortunately, both led to potentially complex weight calculations and overly variable weights, which introduce variance into estimates. New technologies, such as unmanned aerial vehicles, also have the potential to reduce the time and costs involved in listing operations.[23]

Cost is also an important consideration when deciding between traditional and innovative methods. Any non-traditional method will incur additional costs associated with preparation and training, but these will decrease over time as familiarity grows. For example, the DRC two-stage mapping exercise required the purchase of imagery costing $10,000, as well as three weeks of work from an experienced GIS specialist (who was new to this specific image processing application; a specialist with experience in the mapping application could have done the processing in less time). Imagery could also be obtained less expensively, by, for example, using lower resolution images or free OpenStreetMap data, where available.[24] The costs of the new techniques must be weighed against the costs of listing, which increases data collection costs by approximately 25% in each cluster. However, the cost of using either type of methodology is lower than employing a non-probability design, which does not guarantee reliable or representative estimates, regardless of the cost of data collection.

Acknowledgements The authors gratefully acknowledge the comments and contributions of Maximilian Hirn, Siobhan Murray, Utz Pape, and Aude-Sophie Rodella of the World Bank, and Sarchil Qader of Flowminder.

---

[22]For a further discussion, see Facebook Code (2017) and LandScan (2017).

[23]See Eckman et al. (2018).

[24]For further detail and availability, see openstreetmap.org.

References

Bauer, J. J. (2014). "Selection Errors of Random Route Samples." *Sociological Methods & Research* 43 (3): 519–544.

Bauer, J. J. (2016). "Biases in Random Route Surveys." *Journal of Survey Statistics and Methodology* 4: 263–287.

Eckman, S., and B. West. (2016). "Analysis of Data from Stratified and Clustered Surveys." In Wolf, C., Joye, D., Smith, T., and Fu, Y. (Eds.), *Handbook of Survey Methodology*. Thousand Oaks: Sage, 477–487.

Eckman, S., J. Eyerman, and D. Temple. (2018). *Unmanned Aircraft Systems Can Improve Survey Data Collection*. Research Triangle Park, NC: RTI Press. http://goo.gl/eRHKpy.

Facebook Code. (2017). *Open Population Datasets and Open Challenges* [Online]. Available at https://code.facebook.com/posts/596471193873876/open-population-datasets-and-open-challenges/. Accessed 6 November 2017.

Graesser, J., A. Cheriyadat, R. R. Vatsavai, V. Chandola, J. Long, and E. Bright. (2012). "Image Based Characterization of Formal and Informal Neighbourhoods in an Urban Landscape." *IEEE Journal of Selected Topics in Applied Earth Observations and Remote Sensing* 5 (4): 1164–1176.

Grais, R. F., A. M. Rose, and J. P. Guthmann. (2007). "Don't Spin the Pen: Two Alternative Methods for Second-Stage Sampling in Urban Cluster Surveys." *Emerging Themes in Epidemiology* 4 (1): 8.

Grosh M., and J. Muñoz. (1996). *A Manual for Planning and Implementing the Living Standards Measurement Study Survey*. Washington, DC: World Bank.

Himelein, K., S. Eckman, and S. Murray. (2014). "Sampling Nomads: A New Technique for Remote, Hard-to-reach, and Mobile Populations." *Journal of Official Statistics* 30 (2): 191–213.

Himelein, K., S. Eckman, S. Murray, and J. Bauer. (2017). "Alternatives to Full Listing for Second Stage Sampling: Methods and Implications." *Statistical Journal of the IAOS* 33 (3): 701–718.

Hirn, M., and A. S. Rodella. (2017). *WASH Poor in a Water-Rich Country: A Diagnostic of Water, Sanitation, Hygiene, and Poverty in the Democratic Republic of Congo*. WASH Poverty Diagnostic Series. Washington DC: World Bank. Available at https://openknowledge.worldbank.org/handle/10986/27320.

7 Methods of Geo-Spatial Sampling   127

Institut Haïtien de Statistique et d'Informatique (IHSI), Développement Institutions et Mondialisation, and World Bank. (2012). *Enquete sur les Conditions de Vie des Ménages Après Seisme (ECVMAS) 2012.* Available at http://www.ihsi.ht/pdf/ecvmas/ecvmas_metadonnees/0_ECHANTILLON/0_ECVMAS_Plan%20Echantillonnage_28052013.pdf. Accessed 1 November 2017.

Kondo, M. C., K. D. Bream, F. K. Barg, and C. C. Branas. (2014). "A Random Spatial Sampling Method in a Rural Developing Nation." *BMC Public Health* 14 (1): 338.

LandScan. (2017). *LandScan—Data Availability* [Online]. Available at http://web.ornl.gov/sci/landscan/landscan_data_avail.shtml. Accessed 6 November 2017.

Minasny, B., A. B. McBratney, and D. J. Walvoort. (2007). "The Variance Quadtree Algorithm: Use for Spatial Sampling Design." *Computers & Geosciences* 33 (3): 383–392.

Muñoz, J., and W. Langeraar. (2013). *A Census-Independent Sampling Strategy for a Household Survey in Myanmar.* Available at http://winegis.com/images/census-independent-GIS-based-sampling-strategy-for-household-surveys-plan-of-action%20removed.pdf. Accessed 2 November 2017.

OpenStreetMap. (2017). OpenStreetMap [Online]. Available at https://www.openstreetmap.org. Accessed 20 November 2017.

Samet, H. (1984). "The Quadtree and Related Hierarchical Data Structures." *ACM Computing Surveys (CSUR)* 16 (2): 187–260.

Thomson, D. R., F. R. Stevens, N. W. Ruktanonchai, A. J. Tatem, and M. C. Castro. (2017). "GridSample: An R Package to Generate Household Survey Primary Sampling Units (PSUs) from Gridded Population Data." *International Journal of Health Geographics* 16 (1): 25.

The opinions expressed in this chapter are those of the author(s) and do not necessarily reflect the views of the International Bank for Reconstruction and Development/The World Bank, its Board of Directors, or the countries they represent.

Open Access This chapter is licensed under the terms of the Creative Commons Attribution 3.0 IGO license (https://creativecommons.org/licenses/by/3.0/igo/), which permits use, sharing, adaptation, distribution and reproduction in any medium or format, as long as you give appropriate credit to the International Bank for Reconstruction and Development/The World Bank, provide a link to the Creative Commons license and indicate if changes were made.

Any dispute related to the use of the works of the International Bank for Reconstruction and Development/The World Bank that cannot be settled amicably shall be submitted to arbitration pursuant to the UNCITRAL rules. The use of the International Bank for Reconstruction and Development/The World Bank's name for any purpose other than for attribution, and the use of the International Bank for Reconstruction and Development/The World Bank's logo, shall be subject to a separate written license agreement between the International Bank for Reconstruction and Development/The World Bank and the user and is not authorized as part of this CC-IGO license. Note that the link provided above includes additional terms and conditions of the license.

The images or other third party material in this chapter are included in the chapter's Creative Commons license, unless indicated otherwise in a credit line to the material. If material is not included in the chapter's Creative Commons license and your intended use is not permitted by statutory regulation or exceeds the permitted use, you will need to obtain permission directly from the copyright holder.

8

Sampling for Representative Surveys of Displaced Populations

Ana Aguilera, Nandini Krishnan, Juan Muñoz,
Flavio Russo Riva, Dhiraj Sharma and Tara Vishwanath

1 Introduction

As of April 2018, the United Nations High Commissioner for Refugees (UNHCR) reported that an estimated 6.6 million Syrians were internally displaced within the country, and that over 5.6 million Syrians

---

A. Aguilera (✉) · N. Krishnan · D. Sharma · T. Vishwanath
World Bank, Washington, DC, USA
e-mail: aaguileradellano@worldbank.org

N. Krishnan
e-mail: nkrishnan@worldbank.org

T. Vishwanath
e-mail: tvishwanath@worldbank.org

J. Muñoz
Sistemas Integrales, Santiago, Chile
e-mail: juan.munoz@ariel.cl

F. R. Riva
São Paulo School of Administration, São Paulo, Brazil

© International Bank for Reconstruction and Development/The World Bank 2020 **129**
J. Hoogeveen and U. Pape (eds.), *Data Collection in Fragile States*,
https://doi.org/10.1007/978-3-030-25120-8_8

had fled to seek refuge in other countries, of which around 8% were accommodated in camps.[1] In addition to these official figures, there were anywhere from 0.4 to 1.1 million unregistered Syrian refugees in Lebanon and Jordan, and an estimated one million Syrian asylum-seekers in Europe.[2] In effect, more than half of Syria's pre-war population has been forcibly displaced since the beginning of the Syrian civil war.

The Syrian crisis has caused one of the largest episodes of forced displacement since World War II and some of the densest refugee-hosting situations in modern history. Syria's immediate neighbors host the bulk of Syrian refugees: Turkey, Lebanon, and Jordan rank in the top five countries globally for the number of refugees hosted—according to UNHCR data, as of June 2018, Turkey hosted 3.5 million Syrian refugees, Lebanon 0.97 million, and Jordan 0.66 million. In fact, Lebanon and Jordan hold the top two slots for per-capita recipients of refugees in the world, at 164 and 71 refugees per 1000 inhabitants, respectively (UNHCR 2019).[3] The influx into these countries has also occurred at a more rapid rate than prior refugee crises. At one point in the conflict, an average of 6000 Syrians were fleeing into neighboring countries every day.[4] Beyond the immediate impact of inflow of refugees, the host countries are also dealing with other consequences of the

---

[1] http://www.unhcr.org/en-us/syria-emergency.html.

[2] According to a 2014 background paper on Unregistered Syrian Refugees in Lebanon, from the Lebanon Humanitarian INGO Forum, "general estimates and media reports citing unnamed Lebanese officials put the number of Syrians living in Lebanon and not registered with UNHCR between 200,000 and 400,000, although the reliability of and sources for these estimates—which do not distinguish between those in need of protection and/or assistance and those not in need— are unknown" (Lebanon Humanitarian INGO Forum 2014). The paper cites a range of estimates (from around 10 to 50%) based on data from various sources, with differing coverage and survey periods. The 2015 Jordanian census estimated 500,000–600,000 more Syrians than the numbers registered with UNHCR.

[3] Since these figures are based on official UNHCR registration numbers, they do not reflect the unknown number of unregistered refugees, as already noted in footnote 2. At the end of 2014, the United Nations estimated that registered Syrian refugees represented 29% of the total population in Lebanon and 9.5% of the total population in Jordan. Areas with the largest number of Syrians, such as the Bekaa Valley in Lebanon, have seen much higher proportions of refugees to local citizens.

[4] Quoted by the UN High Commissioner for Refugees in a speech to the United Nations Security Council in 2013.

8 Sampling for Representative Surveys of Displaced Populations 131

Syrian conflict, including the disruption on trade and economic activity and growth and spread of the Islamic State (also called ISIS) in Iraq. While the Kurdish Region of Iraq (KRI) hosts at least 200,000 Syrian refugees, the ISIS-induced displacement from neighboring parts of Iraq means that KRI is now hosting over 2.25 million displaced persons, equivalent to approximately 40–50% of its population.

While each neighboring country has received many Syrian refugees in both absolute and relative terms, that is where the commonality ends. Each country has responded to the influx in its own way, influenced by its previous experience of handling protracted displacement situations. Given its history of encampment of the displaced Palestinian population, Lebanon has refrained from setting up camps for Syrians. There is also understandable wariness and anxiety of the impact the influx may have in the delicate domestic political power-sharing equilibrium. In KRI, the influx of Syrian refugees overlaps with a significant number of Iraqi citizens seeking a safe haven from the ISIS militants. The refugees and internally displaced people (IDPs) are located both in camps and non-camps, with a very porous camp boundary that allows its residents to move freely and work outside the camp. At the time of the survey, Jordan had an explicit policy to house refugees in camps and few refugees have legal residency and/or work permits, although a significant majority of refugees had moved outside the camps.

Creating an evidence base to frame the policies for refugees in host environment requires a sampling methodology to select a sample that represents both the host and refugee populations. There are several challenges associated with conducting a representative survey of the host community population and the forcibly displaced. In all three settings we consider, a reliable and updated sampling frame for the resident population was not available.[5] No sample frames existed for forcibly displaced populations as they were excluded from available national sampling frames. Databases maintained by humanitarian agencies for internal programming purposes are often incomplete and out of date.

---

[5]The last official population census in Lebanon was in 1932 and the available sampling frames were also considerably dated in Jordan and KRI.

The displaced also have high degree of mobility and they are often unwilling to speak to surveyors. In this context, and in similar contexts of forced displacement, the selection of a representative sample of hosts and the displaced becomes a major challenge to drawing credible inferences about their socio-economic outcomes.

In this chapter, we describe the strategies that had to be devised to overcome these challenges when designing the sampling procedure for the Syrian Refugee and Host Community Surveys (SRHCS), which were implemented over 2015–2016 in Lebanon, Jordan, and the Kurdistan region of Iraq.[6] Section 2 describes the innovative use of available information to come up with a strategy for generating representative samples of host community and refugee households in the three settings. Section 3 presents the implementation of this strategy. Section 4 concludes by highlighting implementation challenges and drawing general lessons from our experience on sampling forcibly displaced populations.

2   The Innovation

In all three settings, the main challenge to implementing a survey that would yield estimates representative of the refugee and host community populations, was the lack of an updated or comprehensive sample frame, including for hosting populations and especially for displaced populations. In general, the latter were completely missing from existing national sample frames. None of the three countries had at the time, a recent population and housing census, duly updated for population growth and movement, which could have provided the frame to choose the survey sample for the hosting community.

Each of the three contexts presented different challenges. Lebanon and Iraq have both not had a census for several decades and existing sample frames were out of date at the time of the SRHCS. In Lebanon, information from this sample frame was not available at low levels of geographic disaggregation, while in Iraq, internal displacement of

---

[6]The survey was conducted to support analysis on impacts of the influx on local communities in the three settings (see World Bank 2018b).

8 Sampling for Representative Surveys of Displaced Populations   133

millions of Iraqis had made existing frames obsolete. In Jordan, while census exercises are undertaken every decade, data from the most recent census was not available for the SRHCS, and we had to rely on a relatively outdated sample frame based on the 2005 census. Differences in the distribution of Syrian refugees across the three contexts implied a country-specific approach as well. In Lebanon, there were no refugee camps for Syrians; in Jordan, there were two main refugee camps for Syrians; and in Kurdistan, Iraq, Syrians as well as Iraqi IDPs lived in camps but were also free to move in and out.

Defining a sampling strategy to yield representative samples of hosts and displaced populations in this context involved two key innovations. The first was the creation of a sample frame feasible for household listing operations from large geographical divisions where it did not exist. This was the case in Lebanon and among the two largest refugee camps in Jordan. In Lebanon, cartographic divisions of the country were only available for large areas, and had to be segmented and subsegmented based on satellite imagery and dwelling counts to yield geographic areas small enough for listing. These segmentations attempted to divide the larger areas into equal population size subdivisions or segments, much the same way as enumeration areas are generated. Similarly, for the two largest refugee camps in Jordan, Zaatari, and Azraq, satellite imagery was used to divide the camps into mutually exhaustive and exclusive sampling units of roughly equal population size.

The second innovation was the use of available information from different sources on displaced population prevalence which were incorporated into the sample frames of host population prevalence. In most cases, this information was only available at a geographic level higher than the smaller sampling units used in the final frame. This data allowed for the estimation of known probabilities of selection. The first stage sample selection assumed these probabilities were uniformly distributed over the larger geographic area, and in the sampling units within that area. The household listing operation in the selected small sampling units was then used to update this known (albeit incorrect) probability of selection. In Lebanon and Kurdistan, auxiliary information on spatial distribution of refugees and IDPs available from the UNHCR and the International Organization for Migration (IOM), was merged with the sampling frame. Subdistrict level refugee and IDP prevalence information was used to stratify subdistricts by intensity of prevalence: low, middle, and high. The sample was further stratified into subgroups of interest, depending on the context. In Lebanon, the survey was representative of the host community and the Syrian refugee population. In Kurdistan, the scope of the survey was expanded to include IDPs, so that the survey was representative of the host community, Syrian refugees inside and outside of camps, and IDPs inside and outside of camps.

3 Implementation

In what follows, we detail the sampling strategy for Lebanon, which was the most complicated, and then describe the strategy for the other two contexts.

Lebanon. Conducting a representative survey in Lebanon was especially challenging. The first difficulty was that, as of 2015, there was no recent or reliable sample frame, even for Lebanese households, as the last official population census was conducted in 1932. Typically, such a sample frame consists of the universe of enumeration areas in a country, with associated estimates of population. This meant that we had to construct our own sample frame by selecting a few Small Area Units (SAUs) and then conducting a full listing operation by visiting every household within the selected SAUs and collecting basic demographic and contact information. The second difficulty was that there was no available cartographic division of the country into geographic areas small enough to be the subject of a full listing operation, which could then serve as a sampling frame for the SAUs. Circonscription Foncières (CF) were the finest level of disaggregation available; CFs are generally too large to be listed as some have populations of over 100,000. Finally, there was no available sampling frame for Syrian refugees in Lebanon, which meant that we had to depend on UNHCR data on registered Syrian refugees, combined with the estimates of Lebanese population at the CF level. Given these challenges and time and budgetary constraints, the sample was selected in multiple (four) stages as described below.

8 Sampling for Representative Surveys of Displaced Populations

3.1 First Sampling Stage

The sample frame for the first stage is the list of 1301 CFs published by the Council for Development and Reconstruction (CDR) in 2004 and the 2014 UNHCR registration database. Each CF is identified by way of its administrative affiliation—Kaza, Qadha, and Mohafza. The UNHCR database reports the total population in each CF, as well as the number of Lebanese and Syrian population in each.[7,8,9] The CF cartographic boundaries are described digitally in a linked Geographic Information System shape file.

The CFs were sorted into three strata depending on their ex-ante prevalence of Syrian population, as follows:

- Low prevalence: where the Syrian population accounted for less than 20% of the total population;
- Medium prevalence: where the Syrian population accounted for between 20 and 50% of the total population;
- High prevalence: where the Syrian population accounted for over 50% of the total population.

Prevalence of Syrian refugees at the CF level was defined as the number of registered Syrian refugees from the 2014 UNHCR database divided by the sum of the number of registered Syrian refugees and the 2004 Lebanese population counts from the CDR database. The first columns of Table 1 show the distribution of the CFs into strata, as well as the population in each stratum, as per the UNHCR database.

---

[7]Lebanese population distribution by cadasters, supplied by CDR Shapefile (2002–2003); Population estimate of Lebanese 4 million referenced in the Lebanon Crisis Response Plan (LCRP) (UNHCR 2015).

[8]Total population of Syrian refuges as reported by the UNHCR registration database as of December 2014.

[9]Total population of Palestinian refugees in Lebanon (PRL) estimated between 260,000 and 280,000 (UNRWA-AUB 2010). Database provided the population distribution by camps and gatherings. In addition, the total population of Palestinian refugees from Syria is estimated to be 43,000 according to the UNRWA; UNHABITAT UNDP study on gatherings.

136   A. Aguilera et al.

Our intention was to select 75 CFs in total. The decision of how to distribute them across the 3 strata faced the classical dilemma of whether to do it in proportion to the population of the strata, which would deliver nearly optimal estimates for the country as a whole, or to allocate the same sample size (i.e. 25 CFs) to each stratum, which would deliver estimates of nearly the same quality for each of them. Since both considerations were important for the 2015 SRHCS, we opted to do it in accordance to Markwardt's rule (also known as the '50/50 equal/proportional allocation'), which is generally considered a good compromise between the two extremes. The last three columns in Table 1 show the chosen allocation, the corresponding sample sizes (in number of households), and the expected maximum margins of error.[10]

Within each stratum, CFs were selected for inclusion with probability proportional to size (PPS), using the total population as a measure of size, and with implicit stratification by administrative units (Kaza, Qadha and Mohafza). Some of the large CFs were selected more than once. For instance, there were 34 selections made from among the 'low prevalence' CFs (as per Table 1), and one extremely populous CF (Chiyah, located in Mount Lebanon) was randomly selected three times. As a result, the 75 selections were drawn from 71 different CFs. Annex Table 1 shows the list of sampled CFs, where the last column indicates the number of times each CFs was selected in the sample (e.g. one, two or three times depending on each case).

---

[10]More precisely, the last column of Table 1 shows the maximum expected margins of error for the estimation of a household-level prevalence P (such as the percentage of households with children, the percent of households reporting illnesses, etc.) at the 95% confidence level. These are given by $ME = 1.96 \, [\text{Deff} \, P \, (1-P)/n]^{0.5}$, where n is the sample size and Deff is the *design effect*, basically due to the tendency of neighboring households to behave similarly in regards the indicator being observed. The column was computed for $\text{Deff} = 2$ (a value found in practice for many indicators of interest) and $P = 0.5$ (for which ME is maximum).

8 Sampling for Representative Surveys of Displaced Populations 137

3.2 Segmentation of Circonscriptions Foncières (PSUs)

Given that CFs are larger in size than typical census Enumeration Areas which are roughly of 200 households each, the majority of the selected sample CFs was too large to be manageable for implementing a complete household listing operation. For this reason, these large CFs were divided into 'super segments' and 'segments' of roughly equal size within each category, using total number of households as a measure of size. The number of households in each 'super segment' or 'segment' was estimated based on observation of height of buildings and estimated population density in each area in the 2015 ESRI World Imagery[11] and 2015 Google Earth imagery, combined with local knowledge of these areas.

Based on the estimated measure of size, only five CFs were considered to be too large in size and hence were selected for 'super segmentation'. At a later stage, all CFs and 'super segments' were divided into 'segments' due to their large size.

3.3 Second Sampling Stage: Super Segmentation of Circonscriptions Foncières

In the second stage, the boundaries of the 'super segments' in each CF were drawn using the 2015 ESRI World imagery basemap. These boundaries take into account the total estimated household count, as well as natural boundaries such as major roads, rivers, and paths that can easily be recognizable by field teams during the listing operation and implementation of the household questionnaire.

Within each super-segmented CFs, the sample 'super segments' were selected with equal probability, based on the assumption that each 'super segment' is of roughly equal size. The number of 'super segments' selected within each CF was the same as the number of times the corresponding CF was selected in the first sampling stage. For instance, if a

---

[11]Esri, DigitalGlobe, GeoEye, Earthstar Geographics, CNES/Airbus DS, USDA, USGS, AEX, Getmapping, Aerogrid, IGN, IGP, swisstopo, and the GIS User Community.

CF was selected three times in the first sampling stage, we selected three 'super segments' within this CF. Similarly, if a CF was selected only once or twice on the first sampling stage, we correspondingly selected one or two 'super segments' on the secondary sampling stage.

Annex Table 2 shows the list of 'super segments' within selected CFs, where the ninth column indicates the number of times each CFs was selected in the sample (e.g. one, two or three times depending on each case). The column headed 'Prob 2' shows the probability of selecting the 'super segment' within each CF.

3.4 Third Sampling Stage: Segmentation of Circonscriptions Foncières

In a third stage, the boundaries of the 'segments' were drawn for all CFs and selected 'super segments' within CFs. Similar to the process of 'super segmentation', boundaries of segments were drawn using the 2015 ESRI World imagery basemap. These boundaries also take into account the total estimated household count, as well as natural boundaries such as major roads, rivers, and paths.

Within each CF or corresponding 'super segment', the sample 'segments' were selected with equal probability, with the underlying assumption that each 'segment' is of roughly equal size. Annex Table 3 shows the list of 'segments' for all CFs, where the last column indicates the probability of selecting the 'segment' within each CF in the third sampling stage.

3.5 Fourth Sampling Stage

The sample frame for the fourth stage is the full list of all households in the sample CF segments. The listing operation consisted of a full enumeration of all physical structures in the area, with each physical structure being classified as a primary or secondary residential dwelling, commercial building, school, hospital, government office, etc. The listing operation collected information about the household occupying each residential dwelling, and each household was classified as either a Syrian refugee

8 Sampling for Representative Surveys of Displaced Populations

household or a host community household. Care was also taken to record two households living in the same unit separately.[12]

To ensure the quality and completeness of the listing operation, enumerators relied on high-resolution paper maps identifying all buildings within each segment. Each building or structure was pre-assigned with a unique identifier. Enumerators then created a record for each residential unit and household following the protocol described in the 2015 SRHCS Manual of Enumerator. The 40 households to be visited by the 2015 SRHCS in each segment (with a target of 20 Syrian refugee and 20 non-Syrian refugee households in each) was selected from the listing data by systematic equal-probability sampling.[13]

3.6 Selection Probabilities and Sampling Weights

Given the sampling design discussed in the last paragraphs, the probability p_{hizsj} of selecting household hijzsj in segment hizs of super segment hiz in Circonscription Foncière hi of stratum h is given by:

$$p_{hizsj} = \frac{k_h n_{hi}}{\sum_i n_{hi}} \times \frac{t_{hi}}{T_{hi}} \times \frac{g_{hi}}{G_{hi}} \times \frac{m_{hij}}{n'_{hi}}$$

where the four fractions on the right-hand side respectively represent the probability of selecting the CF in the first stage, and the conditional

---

[12]One segment (in the Saida Ed-Dekermane CF, segment number 61119-0-26) was dropped from the original sample since the field team could not get access to the area due to insecurity and was thus unable to implement the household listing operation. Therefore, the intended sample of 40 household in this segment was distributed among two other similar segments, selecting 20 additional households in each. The selection of these two segments was based on the household listing data and local knowledge provided by the survey firm. The two identified segments are located in Saida Al-Qadima and Mazraa 2 (Beirut) and are similar to the Saida Ed-Dekermane segment in that they have: (i) a high share of Palestinian refugees; (ii) high density of urban population; and (iii) high poverty rate.

[13]After listing, only 15 households were found in segment 31116-11. Therefore, all eligible households were selected for interviewing (full census). The total sample size was reduced by 25, for a total 2975 sample households.

probabilities of selecting the super segment, the segment, and the household in the second, third, and fourth stages, and:

* k_h is the number of CFs selected in the stratum (the fifth column in Table 1),
* n_{hi} is the number of households in the CF, as per the sample frame (the column headed 'population' in Table 1),
* t_{hi} is the number of 'super segments' to be drawn in the CF, as per the first sampling stage (the column headed 'No. super segments selected' in Annex Table 2),
* T_{hi} is the total number of 'super segments' in the CF, as per the segmentation procedure (the column headed 'No. of super segments' in Annex Table 2),
* g_{hi} is the number of segments to be drawn in the CF, as per the second sampling stage (the column headed 'n_segments to draw' in Annex Table 3),
* G_{hi} is the total number of segments in the CF, as per the segmentation procedure in the third sampling stage (the column headed 'n_ segments per SSU' in Annex Table 3),
* m_{hij} is the total number of households identified as Syrian refugees during the household listing operation;
* m_{hizsj} is the number of households selected in the segmented CF (with a target 20 Syrian-refugee and 20 non-Syrian-refugee households in this case); or mhij = mhij + (40 − mhij);
* n'hizs is the number of households in the segmented CF, as per the household listing operation.

To deliver unbiased estimates from the sample, the data from each household hij should be affected by a sampling weight (or raising factor) whzsij, equal to the inverse of its selection probability (i.e. whizsj = phizsj − 1).

Kurdistan. Much of the sampling procedure in Kurdistan resembled that of Lebanon, except for one important difference: unlike in Lebanon, the frame for the first stage sample existed in Kurdistan (albeit outdated), and a subset of the enumerations areas had updated population information from the 2012 IHSES survey (which did not take into account subsequent internal displacement). A subsample of the 2012

8 Sampling for Representative Surveys of Displaced Populations 141

clusters was selected for our survey, followed by a comprehensive listing exercise to update the frame for second stage sampling. Four strata based on refugee and IDP prevalence were defined as following:

- Low Syrian prevalence (<5%) and Low IDP prevalence (<15%)
- Low Syrian prevalence (<5%) and High IDP prevalence (>=15%)
- High Syrian prevalence (>=5%) and Low IDP prevalence (<15%).
- High Syrian prevalence (>=5%) and High IDP prevalence (>=15%).

In the first stage, within each stratum, enumeration areas were selected with PPS using the number of households reported from the 2012 listing exercise as a measure of size. In the second stage, 18 households per PSU were selected: six Syrian households, six IDP households, and six host community households in each PSU to the extent possible. In areas where there were less than six Syrian or IDP households, the shortfall was met by host community households. The sampling frame for second stage sampling was the complete list of households in the selected EAs from the listing exercise.

Jordan. In contrast to Lebanon and Iraq, Jordan has carried out Population and Housing Censuses on regular intervals, with the last one in late 2015. What was particularly attractive about the latest census from the perspective of sampling was that it explicitly asked about the nationality of all residents. This would have allowed stratification of areas by density of Syrians. However, the original design could not be implemented because we could not access the new sample frame based on the 2015 Jordanian census. The design was then amended to include a representative sample of the Azraq and Za'atari camps (which account for the vast majority of Syrian refugees in camps in Jordan). This sample was complemented by purposive samples of the surrounding governorates, Mafraq and Zarqa, where the sample included areas physically proximate to the camp and other areas with a high number of Syrian refugees. In Amman Governorate, a purposive sample was drawn, combining a geographically distributed sample with a sample of areas with a high prevalence of Syrian refugees per the 2015 census, as indicated by the Jordanian Department of Statistics. Analytically, this implies the insights from Jordan will be limited to camp residents, neighboring areas of the camps, and Amman governorate.

4 Implementation Challenges, Lessons Learned, and Next Steps

The three surveys described in this paper were designed to generate comparable findings on the lives and livelihoods of Syrian refugees and host communities in the three settings. The absence of updated national sample frames and the lack of a comprehensive mapping of the forced displaced within these countries posed challenges for the design of these surveys. These challenges are not unique—indeed, most developing countries face similar issues, which are exacerbated at times of large scale internal population movements or in contexts of a large localized or widespread influx of migrants. Such data challenges become particularly stark in countries hosting displaced populations or in situations of ongoing or protracted conflict as local populations move to escape violence. But exclusion of displaced persons from national sampling frames, and consequently from national surveys, provides a skewed picture of the world (World Bank 2018a). As the number of displaced persons continues to increase, it becomes all the more urgent to devise strategies to include them in representative socioeconomic surveys.

This methodology paper describes the strategy implemented in the three contexts to generate known ex-ante selection probabilities through a variety of data sources, the use of geospatial segmenting to create enumeration areas where they did not exist, and to use data collected by humanitarian agencies to generate sample frames for displaced populations. The strategies implemented in these surveys can be useful in designing similar exercises in contexts of forced displacement. Moreover, this effort shows the importance of including refugees and non-nationals in national sample frames. The move by Jordan's statistical agency to explicitly include non-nationals in the 2017/2018 household survey is a commendable step in the right direction.

Annex

See Tables 1, 2, and 3.

Table 1 Syrian Refugee and Host Community Survey: sampling strata—Lebanon

Stratum	Prevalence	Sample frame		Syrian Refugee and Host Community Survey 2015		
		No. of CFs	Population	No. of selections	Sample size (HHs)	Margin of error (%)
1. Low prevalence	<= 0.20	946	3,003,958	34	1360	3.76
2. Medium prevalence	0.21–0.50	273	1,039,171	24	960	4.47
3. High prevalence	0.51–1.00	82	465,867	17	680	5.31
Total		1301	4,508,995	75	3000	2.53

Table 2 List of selected segments (enumeration areas)—Lebanon

Segment serial number	CF CAS code	CF name	Qadha name	Mohafza name	Total Syrian population (combined CF)	Total population (combined CF)	No. of polygons	Prevalence of Syrians	Stratum 1-3	Prob 1	Times associated CF selected
1	10210	Msaitbé foncière	Beirut	Beirut	3508	93,838	1	0.04	1	0.98263	1
2	10310	Mazraa foncière	Beirut	Beirut	12,410	125,792	1	0.10	1	1.31724	2
3	10310	Mazraa foncière	Beirut	Beirut	12,410	125,792	1	0.10	1	1.31724	2
4	10650	Achrafieh foncière	Beirut	Beirut	3108	71,541	1	0.04	1	0.74915	1
5	21111	Chiyah	Baabda	Mount Lebanon	50,085	251,061	1	0.20	1	2.62901	3
6	21111	Chiyah	Baabda	Mount Lebanon	50,085	251,061	1	0.20	1	2.62901	3
7	21111	Chiyah	Baabda	Mount Lebanon	50,085	251,061	1	0.20	1	2.62901	3
8	21177	Bourj El-Brajneh	Baabda	Mount Lebanon	24,065	139,404	1	0.17	1	1.45978	2
9	21177	Bourj El-Brajneh	Baabda	Mount Lebanon	24,065	139,404	1	0.17	1	1.45978	2
10	21219	Hadath Beyrouth	Baabda	Mount Lebanon	2702	26,829	1	0.10	1	0.28094	1
11	22111	Bourj Hammoud	El Metn	Mount Lebanon	18,456	94,232	1	0.20	1	0.98676	1
12	22155	Sinn El-Fil	El Metn	Mount Lebanon	3498	38,208	1	0.09	1	0.40010	1
13	22228	Baouchriyé	El Metn	Mount Lebanon	7317	72,611	1	0.10	1	0.76035	1
14	22359	Byaqout	El Metn	Mount Lebanon	346	3753	1	0.09	1	0.03930	1
15	22611	Broummana El-Matn	El Metn	Mount Lebanon	980	8844	1	0.11	1	0.09261	1
16	23469	Aain Zhalta	Chouf	Mount Lebanon	164	1910	1	0.09	1	0.02000	1
17	25111	Jounié Sarba	Kasrouane	Mount Lebanon	775	15,489	1	0.05	1	0.16219	1
18	25211	Aajaltoun	Kasrouane	Mount Lebanon	401	4554	1	0.09	1	0.04769	1
19	26141	Aamchit	Jubail	Mount Lebanon	791	14,288	1	0.06	1	0.14962	1
20	31116	Trablous El-Haddadine	Tripoli	North	1703	53,893	1	0.03	1	0.56435	1
21	31151	Trablous El-Qobbe	Tripoli	North	10,079	65,830	1	0.15	1	0.68935	1
22	32189	Bkeftine	Koura	North	77	881	1	0.09	1	0.00923	1
23	35179	Qboula	Akkar	North	4	616	1	0.01	1	0.00645	1
24	35487	Qbaiyat Aakkar	Akkar	North	568	6973	1	0.08	1	0.07302	1
25	51131	Zahlé Haouch El-Oumara	Zahle	Bekaa	29	5757	1	0.01	1	0.06028	1

(continued)

Table 2 (continued)

Segment serial number	CF CAS code	CF name	Qadha name	Mohafza name	Total Syrian population (combined CF)	Total population (combined CF)	No. of polygons	Prevalence of Syrians	Stratum 1-3	Prob 1	Times associated CF selected
26	53451	Haour Taala	Baalbek	Bekaa	198	3,478	1	0.06	1	0.03642	1
27	61119	Saida Ed-Dekermane	Saida	South	3	60,366	1	0.00	1	0.63213	1
28	61183	Miyé ou Miyé	Saida	South	2453	25,610	1	0.10	1	0.26818	1
29	61489	Aanqoun	Saida	South	645	5386	1	0.12	1	0.05640	1
30	62211	Jouaiya	Sour	South	467	7364	1	0.06	1	0.07711	1
31	62276	Aabbassiyet Sour	Sour	South	2171	14,082	1	0.15	1	0.14746	1
32	71236	Sarba En-Nabatieh	Nabatiye	Nabatiye	68	799	1	0.09	1	0.00837	1
33	72143	Aain Ibl	Bint Jubail	Nabatiye	153	2734	1	0.06	1	0.02863	1
34	74111	Hasbaiya	Hasbaiya	Nabatiye	575	8310	1	0.07	1	0.08702	1
35	22375	Dbayé	El Metn	Mount Lebanon	784	3268	1	0.24	2	0.05969	1
36	23211	Chhim	Chouf	Mount Lebanon	6067	19,616	1	0.31	2	0.35826	1
37	23321	Rmeilet Ech-Chouf	Chouf	Mount Lebanon	2351	4734	1	0.50	2	0.08646	1
38	24111	Choueifat El-Aamrousiyé	Aley	Mount Lebanon	19,572	73,031	1	0.27	2	1.33381	1
39	24133	Choueifat El-Quoubbé	Aley	Mount Lebanon	5843	26,791	1	0.22	2	0.48930	1
40	24343	Bayssour Aaley	Aley	Mount Lebanon	1706	8019	1	0.21	2	0.14646	1
41	31161	Trablous et Tabbaneh	Tripoli	North	6404	26,311	1	0.24	2	0.48053	1
42	32113	Kfar Aaqqa	Koura	North	923	3778	1	0.24	2	0.06900	1
43	33111	Zgharta	Zgharta	North	3218	15,813	1	0.20	2	0.28880	1
44	34269	Aabrine	Batroun	North	447	1753	1	0.25	2	0.03202	1
45	35275	Bebnine	Akkar	North	5301	18,073	1	0.29	2	0.33008	1
46	35364	Ouadi El-Jamous	Akkar	North	1619	5924	1	0.27	2	0.10819	1
47	37231	Beddaoui	Minieh-Danieh	North	16,976	44,404	1	0.38	2	0.81098	1
48	37271	Minie	Minieh-Danieh	North	17,610	38,905	1	0.45	2	0.71054	1
49	51133	Zahlé Aradi	Zahle	Bekaa	1232	6151	1	0.20	2	0.11234	1
50	51224	Jdita	Zahle	Bekaa	2990	9242	1	0.32	2	0.16879	1

(continued)

Table 2   (continued)

Segment serial number	CF CAS code	CF name	Qadha name	Mohafza name	Total Syrian population (combined CF)	Total population (combined CF)	No. of polygons	Prevalence of Syrians	Stratum 1-3	Prob 1	Times associated CF selected
51	52224	Baaloul BG	West Bekaa	Bekaa	871	2089	1	0.42	2	0.03815	1
52	53111	Baalbek	Baalbek	Bekaa	22,898	71,504	1	0.32	2	1.30592	1
53	53167	Saaidé	Baalbek	Bekaa	761	1647	1	0.46	2	0.03008	1
54	53311	Deir El-Ahmar	Baalbek	Bekaa	2924	7442	1	0.39	2	0.13592	1
55	53445	Nabi Chit	Baalbek	Bekaa	3094	9603	1	0.32	2	0.17539	1
56	61311	Ghaziyé	Saida	South	5163	18,290	1	0.28	2	0.33404	1
57	71113	Nabatiyeh El-Faouka	Nabatiye	Nabatiye	2568	6905	1	0.37	2	0.12611	1
58	74122	Hebbariyé	Hasbaiya	Nabatiye	780	2484	1	0.31	2	0.04537	1
59	24211	Aaramoun Aaley	Aley	Mount Lebanon	9827	15,666	1	0.63	3	0.50870	1
60	31111	Trablous Ez-Zeitoun	Tripoli	North	18,633	23,529	1	0.79	3	0.76402	1
61	35111	Halba	Akkar	North	10,842	16,668	1	0.65	3	0.54123	1
62	35429	Kouachra	Akkar	North	1958	3177	1	0.62	3	0.10316	1
63	35516	Mazareaa Jabal Akroum	Akkar	North	5965	11,487	1	0.52	3	0.37300	1
64	37317	Bqaa Sefrine	Minieh-Danieh	North	2224	4271	1	0.52	3	0.13869	1
65	51125	Zahlé Maallaqa Aradi	Zahle	Bekaa	6171	10,097	1	0.61	3	0.32786	1
66	51231	Saadnayel	Zahle	Bekaa	16,293	23,393	1	0.70	3	0.75961	1
67	51234	Qabb Elias	Zahle	Bekaa	27,951	39,206	1	0.71	3	1.27308	1
68	51267	Barr Elias	Zahle	Bekaa	34,688	45,306	1	0.77	3	1.47115	1
69	51284	Majdel Aanjar	Zahle	Bekaa	16,722	24,653	1	0.68	3	0.80052	1
70	51311	Riyaq	Zahle	Bekaa	6921	10,808	1	0.64	3	0.35095	1
71	52211	Joubb Jannine	West Bekaa	Bekaa	7833	13,478	1	0.58	3	0.43765	1
72	52234	Khiara	West Bekaa	Bekaa	1577	2004	1	0.79	3	0.06507	1
73	52277	Marj BG	West Bekaa	Bekaa	15,071	18,366	1	0.82	3	0.59637	1
74	61115	Saida El-Qadimeh	Saida	South	14,641	23,658	1	0.62	3	0.76821	1
75	61453	Bissariye	Saida	South	4931	8661	1	0.57	3	0.28124	1

Table 3 List of sample super segments (for CFs divided into super-segments or secondary sampling units)—Lebanon

SN	CAS_code	CF_name	Qadha_name	Mohafza_Na	Total_popu	Super segment ID	Segment ID	n_segments per SSU	n_segments to draw	Rand (TSU)	Prob 3
1	10210	Msaitbé foncière	Beirut	Beirut	93838	10210-7	10210-7-13	18	1	0.02851	0.05556
2	10310	Mazraa foncière	Beirut	Beirut	125792	10310-1	10310-1-18	26	1	0.01869	0.03846
2	10310	Mazraa foncière	Beirut	Beirut	125792	10310-7	10310-7-6	17	1	0.08653	0.05882
4	10650	Achrafieh foncière	Beirut	Beirut	71541	10650-0	10650-0-66	93	1	0.00334	0.01075
5	21111	Chiyah	Baabda	Mount Lebanon	251061	21111-10	21111-10-34	41	1	0.02708	0.02439
5	21111	Chiyah	Baabda	Mount Lebanon	251061	21111-5	21111-5-9	23	1	0.04097	0.04348
5	21111	Chiyah	Baabda	Mount Lebanon	251061	21111-7	21111-7-19	22	1	0.08325	0.04545
8	21177	Bourj El-Brajneh	Baabda	Mount Lebanon	139404	21177-11	21177-11-1	14	1	0.03035	0.07143
8	21177	Bourj El-Brajneh	Baabda	Mount Lebanon	139404	21177-2	21177-2-9	23	1	0.00106	0.04348
10	21219	Hadath Beyrouth	Baabda	Mount Lebanon	26829	21219-0	21219-0-6	28	1	0.10421	0.03571
11	22111	Bourj Hammoud	El Metn	Mount Lebanon	94232	22111-6	22111-6-3	21	1	0.00019	0.04762
12	22155	Sinn El-Fil	El Metn	Mount Lebanon	38208	22155-0	22155-0-66	68	1	0.00901	0.01471
13	22228	Baouchriyé	El Metn	Mount Lebanon	72611	22228-0	22228-0-49	83	1	0.02951	0.01205
14	22359	Byaqout	El Metn	Mount Lebanon	3753	22359-0	22359-0-2	6	1	0.07392	0.16667
35	22375	Dbayé	El Metn	Mount Lebanon	3268	22375-0	22375-0-4	4	1	0.21483	0.25000
15	22611	Broummana El-Matn	El Metn	Mount Lebanon	8844	22611-0	22611-0-2	10	1	0.22362	0.10000
36	23211	Chhim	Chouf	Mount Lebanon	19616	23211-0	23211-0-5	21	1	0.09593	0.04762
37	23321	Rmeilet Ech-Chouf	Chouf	Mount Lebanon	4734	23321-0	23321-0-2	5	1	0.67365	0.20000
16	23469	Aain Zhalta	Chouf	Mount Lebanon	1910	23469-0	23469-0-1	2	1	0.47936	0.50000
38	24111	Choueifat El-Aamrousiyé	Aley	Mount Lebanon	73031	24111-0	24111-0-101	102	1	0.00238	0.00980
39	24133	Choueifat El-Quoubbé	Aley	Mount Lebanon	26791	24133-0	24133-0-11	29	1	0.09931	0.03448
59	24211	Aaramoun Aaley	Aley	Mount Lebanon	15666	24211-0	24211-0-11	18	1	0.06641	0.05556
40	24343	Bayssour Aaley	Aley	Mount Lebanon	8019	24343-0	24343-0-7	10	1	0.02895	0.10000
17	25111	Jounié Sarba	Kasrouane	Mount Lebanon	15489	25111-0	25111-0-20	22	1	0.05377	0.04545
18	25211	Aajaltoun	Kasrouane	Mount Lebanon	4554	25211-0	25211-0-1	5	1	0.09509	0.20000

(continued)

Table 3 (continued)

NS	CAS_code	CF_name	Qadha_name	Mohafza_Na	Total_popu	Super segment ID	Segment ID	n_segments per SSU	n_segments to draw	Rand (TSU)	Prob 3
19	26141	Aamchit	Jubail	Mount Lebanon	14288	26141-0	26141-0-9	14	1	0.10108	0.07143
60	31111	Trablous Ez-Zeitoun	Tripoli	North	23529	31111-0	31111-0-13	48	1	0.01400	0.02083
20	31116	Trablous El-Haddadine	Tripoli	North	53893	31116-0	31116-0-11	54	1	0.01494	0.01852
21	31151	Trablous El-Qobbe	Tripoli	North	65830	31151-0	31151-0-42	44	1	0.00794	0.02273
41	31161	Trablous et Tabbaneh	Tripoli	North	26311	31161-0	31161-0-16	27	1	0.08705	0.03704
42	32113	Kfar Aaqqa	Koura	North	3778	32113-0	32113-0-1	4	1	0.10281	0.25000
22	32189	Bkeftine	Koura	North	881	32189-0	32189-0-1	1	1	0.45403	1.00000
43	33111	Zgharta	Zgharta	North	15813	33111-0	33111-0-9	18	1	0.06386	0.05556
44	34269	Aabrine	Batroun	North	1753	34269-0	34269-0-1	3	1	0.08812	0.33333
61	35111	Halba	Akkar	North	16668	35111-0	35111-0-15	19	1	0.02170	0.05263
23	35179	Qboula	Akkar	North	616	35179-0	35179-0-1	1	1	0.81850	1.00000
45	35275	Bebnine	Akkar	North	18073	35275-0	35275-0-3	21	1	0.04383	0.04762
46	35364	Ouadi El-Jamous	Akkar	North	5924	35364-0	35364-0-9	9	1	0.35237	0.11111
62	35429	Kouachra	Akkar	North	3177	35429-0	35429-0-3	3	1	0.22822	0.33333
24	35487	Qbaiyat Aakkar	Akkar	North	6973	35487-0	35487-0-4	7	1	0.01762	0.14286
63	35516	Mazareaa Jabal Akroum	Akkar	North	11487	35516-0	35516-0-5	11	1	0.18676	0.09091
47	37231	Beddaoui	Minieh-Danieh	North	44404	37231-0	37231-0-50	57	1	0.02521	0.01754
48	37271	Minie	Minieh-Danieh	North	38905	37271-0	37271-0-20	40	1	0.01934	0.02500
64	37317	Bqaa Sefrine	Minieh-Danieh	North	4271	37317-0	37317-0-4	4	1	0.44794	0.25000
65	51125	Zahlé Maallaqa Aradi	Zahle	Bekaa	10097	51125-0	51125-0-4	15	1	0.19174	0.06667
25	51131	Zahlé Haouch El-Oumara	Zahle	Bekaa	5757	51131-0	51131-0-4	6	1	0.12081	0.16667
49	51133	Zahlé Aradi	Zahle	Bekaa	6151	51133-0	51133-0-5	7	1	0.01805	0.14286
50	51224	Jdita	Zahle	Bekaa	9242	51224-0	51224-0-3	11	1	0.01322	0.09091
66	51231	Saadnayel	Zahle	Bekaa	23393	51231-0	51231-0-16	26	1	0.10708	0.03846
67	51234	Qabb Elias	Zahle	Bekaa	39206	51234-0	51234-0-26	35	1	0.00073	0.02857

(continued)

8 Sampling for Representative Surveys of Displaced Populations 149

Table 3 (continued)

SN	CAS_code	CF_name	Qadha_name	Mohafza_Na	Total_popu	Super segment ID	Segment ID	n-segments per SSU	n-segments to draw	Rand (TSU)	Prob 3
68	51267	Barr Elias	Zahle	Bekaa	45306	51267-0	51267-0-14	48	1	0.01760	0.02083
69	51284	Majdel Aanjar	Zahle	Bekaa	24653	51284-0	51284-0-13	25	1	0.01400	0.04000
70	51311	Riyaq	Zahle	Bekaa	10808	51311-0	51311-0-2	11	1	0.07445	0.09091
71	52211	Joubb Jannine	West Bekaa	Bekaa	13478	52211-0	52211-0-1	14	1	0.01374	0.07143
51	52224	Baaloul BG	West Bekaa	Bekaa	2089	52224-0	52224-0-1	2	1	0.19555	0.50000
72	52234	Khiara	West Bekaa	Bekaa	2004	52234-0	52234-0-2	2	1	0.61762	0.50000
73	52277	Marj BG	West Bekaa	Bekaa	18366	52277-0	52277-0-8	20	1	0.13774	0.05000
52	53111	Baalbek	Baalbek	Bekaa	71504	53111-0	53111-0-70	80	1	0.01073	0.01250
53	53167	Saaidé	Baalbek	Bekaa	1647	53167-0	53167-0-1	2	1	0.57735	0.50000
54	53311	Deir El-Ahmar	Baalbek	Bekaa	7442	53311-0	53311-0-5	9	1	0.16490	0.11111
55	53445	Nabi Chit	Baalbek	Bekaa	9603	53445-0	53445-0-10	10	1	0.24514	0.10000
26	53451	Haour Taala	Baalbek	Bekaa	3478	53451-0	53451-0-2	3	1	0.23547	0.33333
74	61115	Saida El-Qadimeh	Saida	South	23658	61115-0	61115-0-16	25	1	0.08783	0.04000
27	61119	Saida Ed-Dekermane	Saida	South	60366	61119-0	61119-0-26	69	1	0.01328	0.01449
28	61183	Miyé ou Miyé	Saida	South	25610	61183-0	61183-0-1	29	1	0.10490	0.03448
56	61311	Ghaziyé	Saida	South	18290	61311-0	61311-0-5	19	1	0.00795	0.05263
75	61453	Bissariye	Saida	South	8661	61453-0	61453-0-6	9	1	0.10027	0.11111
29	61489	Aanqoun	Saida	South	5386	61489-0	61489-0-3	5	1	0.19827	0.20000
30	62211	Jouaiya	Sour	South	7364	62211-0	62211-0-4	9	1	0.20830	0.11111
31	62276	Aabbassiyet Sour	Sour	South	14082	62276-0	62276-0-1	18	1	0.00890	0.05556
57	71113	Nabatiyeh El-Faouka	Nabatiye	Nabatiye	6905	71113-0	71113-0-2	9	1	0.18614	0.11111
32	71236	Sarba En-Nabatieh	Nabatiye	Nabatiye	799	71236-0	71236-0-1	1	1	0.59953	1.00000
33	72143	Aain Ibl	Bint Jubail	Nabatiye	2734	72143-0	72143-0-1	3	1	0.32534	0.33333
34	74111	Hasbaiya	Hasbaiya	Nabatiye	8310	74111-0	74111-0-2	8	1	0.04804	0.12500
58	74122	Hebbariyé	Hasbaiya	Nabatiye	2484	74122-0	74122-0-1	3	1	0.06554	0.33333

References

Lebanon Humanitarian INGO Forum. (2014). *Background Paper on Unregistered Syrian Refugees in Lebanon.* Available at http://lhif.org/uploaded/News/d92fe3a1b1dd46f2a281254fa551bd09LHIF%20Background%20Paper%20on%20Unregistered%20Syrian%20Refugees%20(FINAL).pdf.

The Data Blog. (2017). *A First Look at Facebook's High-Resolution Population Maps* [Online]. Available at http://blogs.worldbank.org/opendata/first-look-facebook-s-high-resolution-population-maps. Accessed 6 November 2017.

UNHCR. (2015). *Lebanon Crisis Response Plan 2015–16.*

UNHCR. (2019). *Global Trends: Forced Displacement in 2018.*

UNRWA-AUB. (2010). *UNRWA-AUB Socio-Economic Survey of Palestine Refugees in Lebanon.* https://www.unrwa.org/newsroom/press-releases/unrwa-aub-socio-economic-survey-palestine-refugees-lebanon.

World Bank. (2018a). *Poverty and Shared Prosperity Report: Piecing Together the Poverty Puzzle.* Washington, DC.

World Bank. (2018b). *Syrian Refugees and Their Hosts in Jordan, Lebanon and the Kurdistan Region of Iraq: Lives, Livelihoods, and Local Impacts.* Unpublished Manuscript.

8 Sampling for Representative Surveys of Displaced Populations

The opinions expressed in this chapter are those of the author(s) and do not necessarily reflect the views of the International Bank for Reconstruction and Development/The World Bank, its Board of Directors, or the countries they represent.

Open Access This chapter is licensed under the terms of the Creative Commons Attribution 3.0 IGO license (https://creativecommons.org/licenses/by/3.0/igo/), which permits use, sharing, adaptation, distribution and reproduction in any medium or format, as long as you give appropriate credit to the International Bank for Reconstruction and Development/The World Bank, provide a link to the Creative Commons license and indicate if changes were made.

Any dispute related to the use of the works of the International Bank for Reconstruction and Development/The World Bank that cannot be settled amicably shall be submitted to arbitration pursuant to the UNCITRAL rules. The use of the International Bank for Reconstruction and Development/The World Bank's name for any purpose other than for attribution, and the use of the International Bank for Reconstruction and Development/The World Bank's logo, shall be subject to a separate written license agreement between the International Bank for Reconstruction and Development/The World Bank and the user and is not authorized as part of this CC-IGO license. Note that the link provided above includes additional terms and conditions of the license.

The images or other third party material in this chapter are included in the chapter's Creative Commons license, unless indicated otherwise in a credit line to the material. If material is not included in the chapter's Creative Commons license and your intended use is not permitted by statutory regulation or exceeds the permitted use, you will need to obtain permission directly from the copyright holder.

9

Rapid Consumption Surveys

Utz Pape and Johan Mistiaen

1 The Data Demand and Challenge

Poverty is the paramount indicator used to gauge the socioeconomic well-being of a population. Particularly after a shock or in a volatile context, poverty estimates can identify who was affected, and how severely. This is particularly relevant in fragile countries where monitoring poverty dynamics help measure the country's progress toward stability, or increased risk of relapsing into conflict. As one of the main indicators for poverty, monetary poverty is measured by a welfare

---

This chapter is a summary of Pape, Utz Johann, and Johan Mistiaen. "Household Expenditure and Poverty Measures in 60 Minutes: A New Approach with Results from Mogadishu." Policy Research Working Paper Series. The World Bank, 2018. https://ideas.repec.org/p/wbk/wbrwps/8430.html.

U. Pape (✉) · J. Mistiaen
World Bank, Washington, DC, USA
e-mail: upape@worldbank.org

J. Mistiaen
e-mail: jmistiaen@worldbank.org

© International Bank for Reconstruction and Development/The World Bank 2020 **153**
J. Hoogeveen and U. Pape (eds.), *Data Collection in Fragile States*,
https://doi.org/10.1007/978-3-030-25120-8_9

aggregate, usually based on consumption in developing countries and a poverty line. The poverty line indicates the minimum level of welfare required for healthy living.

Consumption aggregates are traditionally estimated based on time-consuming household consumption surveys. A household consumption questionnaire records consumption (how much was consumed) and expenditure (how much was purchased, or obtained in other ways like gifts or aid) for a comprehensive list of food and non-food items. Covering between 300 and 400 items, the questionnaire often exceeds 120 minutes to administer. In addition to the longer administering time leading to higher costs, response fatigue can increase measurement error, especially for items at the end of the questionnaire. In a fragile country context, a face-to-face time of 90–120 minutes can be prohibitively high. In the case of Somalia, security concerns restricted the duration of a survey visit in Mogadishu to about 60 minutes.

The extensive nature of household consumption surveys makes it difficult to obtain updated poverty estimates, especially when they are needed the most, such as after a shock and in fragile countries. Approaches have therefore been developed to reduce administering times to allow for the collection of consumption data. The most straightforward approach to minimize administering time is to reduce the number of items surveyed, either by asking for aggregates, or by skipping less frequently consumed items, which is called the reduced consumption methodology. However, both approaches—using aggregates, and skipping less common items–have been shown to underestimate consumption, which in turn overestimates poverty.[1] Splitting the questionnaire to allow for multiple visits is another solution, but potential attrition issues especially in fragile contexts increases the required sample size and may be costlier. In addition, multiple visits to the same household can increase security concerns.

The second class of approaches utilizes a full consumption baseline survey and updates poverty estimates based on a small subset of collected

---

[1]Beegle et al. (2012).

indicators.[2] These approaches estimate a welfare model based on the baseline survey using a small number of easy-to-collect indicators. This allows poverty estimates to be updated by collecting only the set of indicators instead of the direct consumption data. While this approach is cost-effective and easy to implement in normal circumstances, it has two major drawbacks in the context of fragility and shocks. First, the approach requires a baseline survey, which is sometimes not available, as in the case of Mogadishu. Second, the approach relies on a structural model estimated from the baseline survey.[3] In the case of shocks, structural assumptions that cannot be tested are often violated. Thus, poverty updates based on the violated assumptions tend to underestimate the impact of the shock on poverty. Therefore, cross-survey imputation methodologies are not applicable in the context of shocks and fragility. .

2 The Innovation

To assess poverty in Mogadishu, we tested a new methodology combining an innovative questionnaire design with standard imputation techniques. This substantially reduces the administering time of a consumption survey from multiple hours or even days to about 60 minutes, while still resulting incredible poverty estimates. The gain in shorter administering time, however, is offset by the need to impute missing consumption values. Given the design of the questionnaire, this method circumvents the systematic biases identified for alternative methodologies.

2.1 Overview

The rapid consumption survey methodology involves five main steps (Fig. 1). First, core items are selected based on their importance for consumption. Second, the remaining items are partitioned into optional

---

[2]Douidich et al. (2013); SWIFT.
[3]Christiaensen et al. (2011).

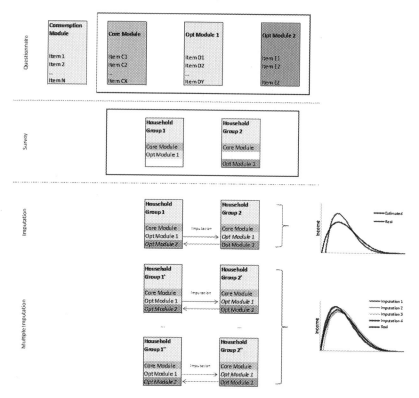

Fig. 1 Illustration of the rapid consumption survey methodology (using illustrative data only)

modules. Third, optional modules are assigned to groups of households. Fourth, after data collection, consumption of optional modules is imputed for all households. Fifth, the resulting consumption aggregate is used to estimate poverty indicators.

First, core consumption items are selected. Consumption in a country bears some variability, but usually a small number of a few dozen items captures the majority of consumption. These items are assigned to the core module, which will be administered to all households. Important items can be identified by their average consumption share

9 Rapid Consumption Surveys   157

per household or across households. Previous consumption surveys in the same country, or consumption shares of neighboring or similar countries can be used to estimate consumption shares.

Second, non-core items are partitioned into optional modules. Different methods can be used for this partitioning. In the simplest case, the remaining items are ordered according to their consumption share and assigned one by one while iterating the optional module in each step. A more sophisticated method takes into account the correlation between items, and partition them in a way so that all items within a module explain consumption as best as possible, while the information between modules should be highly correlated. The partitioning influences the standard error of the estimation, but does not introduce bias. Thus, even in the absence of a previous survey, this methodology can be applied. More complicated partition patterns can result in a set of very different items in each module. However, the modular structure should not influence the layout of the questionnaire. Instead, all items should be grouped into categories of consumption (e.g. cereals) and different recall periods. It is therefore recommended to use CAPI technology, which allows the structure of the consumption module to be hidden from the enumerator.

Third, optional modules should be assigned to groups of households. Optional modules should be assigned randomly, stratified by clusters to ensure appropriate representation of optional modules in each cluster. This means that each cluster should include about the same number of households assigned to each optional module. This step is followed by the actual data collection.

Fourth, household consumption should be estimated by imputation. The average consumption of each optional module can be estimated based on the subsample of households assigned to the optional module. In the most straightforward case, a simple average can be estimated. More sophisticated techniques can employ a welfare model based on household characteristics and consumption of the core items. The next section presents six techniques and demonstrates their performance on the dataset from Hargeisa.

Single imputation of the consumption aggregate underestimates the variance of household consumption. Depending on the location of the poverty line relative to the consumption distribution, this may either consistently under- or overestimate poverty. Multiple imputations based on bootstrapping can mitigate the problem but will render analysis more complicated. We use single as well as multiple imputation techniques for the evaluation of the methodology.

3 Key Results

In this section, the rapid consumption methodology will first be applied to a dataset including a full consumption module from Hargeisa, Somaliland. This will be used to assess the performance of the rapid consumption methodology compared to the traditional full consumption methodology. The results of the High Frequency Survey in Mogadishu are then presented. Security risks in Mogadishu restrict face-to-face interview time to less than one hour; therefore, the rapid consumption methodology was used to derive the first ever consumption estimates for Mogadishu. We present the resulting consumption aggregate, and perform consistency checks for its validation.

3.1 Ex Post Simulation

The rapid consumption methodology is applied ex post to household budget data collected in Hargeisa, Somaliland. Hargeisa was chosen as it is very similar to Mogadishu. Using the full consumption dataset from Hargeisa allows a full assessment of the new methodology. Based on selected indicators, we compare the results of the estimated consumption based on the rapid consumption methodology with the results from using the traditional full consumption module. We add a comparison with the results for a reduced consumption module.

The simulation assigns each household to one optional module. The consumption data for the modules not assigned to the household is deleted. Multiple simulations are performed, with various modules being assigned to households. Across the simulations, we calculate three consumption indicators and four poverty and inequality indicators. The

9 Rapid Consumption Surveys   159

Table 1 Number of items and consumption share captured per module

	Number of food items	Share of con-sumption (%)	Number of non-food items	Share of con-sumption (%)
Core	33	92	25	88
Module 1	17	3	15	3
Module 2	17	2	15	3
Module 3	15	2	15	4
Module 4	17	2	15	3

consumption indicators capture the accuracy of the estimation at three different levels: the household level, the cluster level (consisting of about nine households), and the level of the dataset. In addition, we calculate the poverty headcount (FGT0), the poverty depth (FGT1), the poverty severity (FGT2), and the Gini coefficient to capture inequality.

Six estimation techniques are compared with respect to their relative bias and relative standard error, based on 20 simulations. All simulations used the same item assignment to modules using the algorithm as described (see Table 1 for the resulting consumption shares per module).[4] The estimation techniques differ considerably in terms of performance. We also compare the techniques to using a reduced consumption module where the same consumption items are collected for all households. The number of items is equal to the size of the core module and one optional module, implying a comparable face-to-face interview time to the rapid consumption methodology.

Comparing the reduced consumption approach with the full consumption as a reference, the reduced consumption approach suffers from an underestimation of consumption. This is not surprising because the approach only collects information on the consumption of a subset of items. Applying the median as a summary statistic also results in an underestimation of consumption. As consumption distributions have a long right tail, the median consumption belongs to a poorer household than the average household. In the case of Hargeisa, several optional

---

[4]We performed robustness checks with different item assignment to modules, including setting the parameter $d = 1$ and $d = 2$. The estimation results are extremely robust to changes in the item assignment to modules.

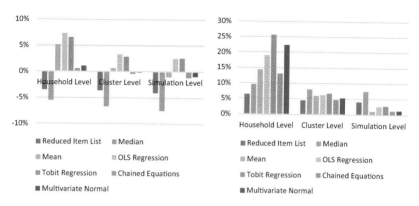

Fig. 2 Average relative bias and standard error

modules have a median of zero consumption. Thus, the median underestimates the consumption in a similar way to the reduced consumption approach. In contrast, the average consumption of households is larger than the consumption of the median household. Thus, it is not surprising that the technique using the average as a summary statistic overestimates total consumption at the household and cluster levels.

The regression techniques have a similar performance, with a considerable upward bias at all levels. The Tobit regression performs slightly better at the household and cluster levels. As known from literature about small area estimates, the regression approaches do not model the error distribution correctly and, thus, underestimate the tails of the distribution. Depending on the value of the poverty line relative to the mode of the distribution, this results in an over- or under-estimation of the poverty rate. In contrast, both imputation techniques perform exceptionally well, with a bias below 1% at all levels (Fig. 2).

While the bias is important in order to understand the systematic deviation of the estimation, the relative standard error helps to understand the variation of the estimation. Other than in a simulation setting, the standard error of the estimation cannot be calculated, as only one assignment of households to optional modules is available. Thus, it is important that the estimation technique delivers a small relative standard error.

Generally, the relative standard error reduces when moving from the household level over the cluster level to the simulation level. The relative standard error for the reduced consumption methodology is smaller than for the summary statistic techniques because the reduced consumption is not subject to variation from the module assignment to households. The regression techniques have large relative standard errors of around 20% at the household level, while the multiple imputation techniques vary between 15 and 20%. At the cluster level, the relative standard error drops to 7% for regression techniques and 5% for multiple imputation techniques. At the simulation level, the relative standard error is around 3% for regression techniques and 1% for multiple imputation techniques.

The distributional shape of the estimated household consumption level can be compared to the reference household consumption by employing standard poverty and inequality indicators. The poverty headcount (FGT0) is 57.4% for the reference distribution.[5] Not surprisingly, the reduced consumption technique and the median summary statistic overestimate poverty by several percentage points due to the underestimation of consumption, while the average summary statistic and the regression techniques underestimate poverty, since they overestimate consumption. The multiple imputation techniques overestimate poverty, but only by 0.5 percentage points (or about 1%), performing significantly better than the reduced consumption approach, which has a bias that is more than two times larger. The reduced consumption technique and the median summary statistic as well as the multiple imputation techniques deliver good results for FGT1 and FGT2, emphasizing that not only can the headcount be estimated reasonably well, but the distributional shape is also conserved. With the exception of the median summary statistic, these techniques also perform well estimating the Gini coefficient, with a bias of less than 0.5 percentage points. The relative standard errors show similar results as for the estimation of the consumption. The relative standard error of the reduced

---

[5]The FGT0 is calculated based on the US$1.90 PPP (2011) international poverty line, converted into local currency in 2013.

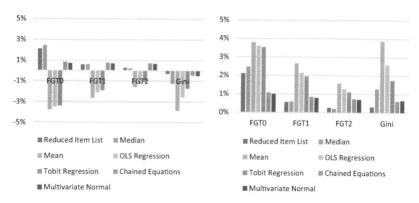

Fig. 3 Bias and standard errors

consumption for FGT0 is double that of the multiple imputation techniques. The relative standard errors for the multiple imputation techniques for FGT1 are comparable but larger than for FGT2 and Gini (Fig. 3).

In conclusion, the average summary statistic and the regression approaches cannot deliver convincing estimates. While the reduced consumption technique and the median summary statistic perform considerably better, they both overestimate poverty. Only the multiple imputation techniques are convincing in all estimation exercises. In terms of the estimation of the important poverty headcount (FGT0), the multiple imputation techniques are virtually unbiased.

4 Implementation Challenges, Lessons Learned, and Next Steps

In late 2014, consumption data using the proposed rapid consumption methodology was collected in Mogadishu using CAPI. The rapid consumption questionnaire reduced face-to-face interview time considerably. A household visit took about 40 minutes on average (with a median of 35 minutes), including greetings, household characteristics, consumption modules, and a number of perception questions. Nine out of ten interviews took less than 65 minutes.

After data cleaning and quality assurance procedures, 675 households with consumption data were retained.[6] A welfare model was built to predict missing consumption in optional modules. The welfare model was tested on the core consumption, after removing the core consumption as an explanatory variable. The model for food consumption retrieved an R2 of 0.24, while non-food consumption was modeled with an R2 of 0.16. It is important to emphasize that these models give a lower bound of the R2 compared to the models used in the prediction, as the prediction models include the core consumption as an explanatory variable. Given the assessment of the different estimation techniques in the previous section, the multivariate normal approximation using multiple imputations is applied to the Mogadishu dataset.

For the Mogadishu dataset, the assignment of items to modules had to be manually refined.[7] The refinement had a minor impact on the share of consumption per module. It is curious, though, that the share of consumption per module is different for Hargeisa and Mogadishu. Using the Hargeisa dataset, 91% of food consumption (and 76% of non-food consumption) is captured in the core module. In contrast, the core food consumption share is only 64% (and 62% of non-food consumption) in Mogadishu before imputing the consumption of non-assigned modules. Thus, employing a reduced consumption module based on consumption shares identified in Hargeisa would have crudely underestimated consumption in Mogadishu, without being able to evaluate the inaccuracy. In contrast, the rapid consumption methodology allows the estimation of shares for each module, while the consumption

---

[6]While the survey also covered IDP camps, the analysis presented is restricted to households in residential areas, excluding IDP camps.

[7]Manual refinement is necessary to ensure that items like 'other fruits' do not double-count types of fruits not assigned to the household. This is implemented by relabeling and manually assigning modules. In addition, some item groups items were split into individual items, which is generally preferable for recall and recording, as well as calculation of unit values.

estimation procedure implicitly takes into account the 'missing' consumption shares for each household (Table 2).

The cumulative consumption distribution can be compared for the consumption captured in the core module, the assigned optional modules, and the imputed consumption. By construction, the core consumption shows the lowest consumption per household. Adding the consumption from the assigned optional modules shifts the cumulative consumption curve slightly. The imputed consumption is shifted even further as the estimated consumption shares from the non-assigned modules are added (Fig. 4).

Without full consumption aggregate values for Mogadishu, we can only show the consistency of the retrieved consumption aggregate with other household characteristics to validate the estimates. Consumption per capita usually reduces with increasing household size. Indeed, we find that household size is significantly negatively correlated with estimated per capita consumption.[8] Per capita consumption also decreases with a larger share of children among the household members. The proportion of employed members of the household significantly increases consumption per capita. Thus, the retrieved consumption estimate is consistent and using the evidence from the ex post simulations, highly accurate.

The results of the ex post simulation indicate that the rapid consumption methodology can reliably estimate consumption and poverty. The experience in Mogadishu also shows that the rapid consumption methodology can be implemented in extremely high-risk areas, due to its success in limiting face-to-face interview time to less than one hour. While these results are encouraging, the rapid consumption methodology has some limitations.

The rapid consumption questionnaire varies in comprehensiveness and the order of items in the consumption module between households.

---

[8]The reported numbers are corrected against correlation with household characteristics included in the welfare model. As the welfare model for the prediction of consumption includes household size, we have run a robustness check excluding household size from the welfare model used for prediction. The correlation between consumption per capita and household size is still significant (coefficient: -0.03, t-statistic: -2.17, p-value: 0.03).

Table 2 Number of items and consumption shares captured per module

	Food consumption				Non-food consumption			
	Number of items	Share Hargeisa (%)	Share Mogadishu (%)	Share Mogadishu imputed (%)	Number of items	Share Hargeisa (%)	Share Mogadishu (%)	Share Mogadishu imputed (%)
Core	33	91	64	54	26	76	62	52
Module 1	19	3	9	16	15	7	9	12
Module 2	20	2	14	14	15	5	9	12
Module 3	15	2	5	6	15	6	8	9
Module 4	15	2	8	9	15	6	11	15

Note simulated for Hargeisa, estimated for Mogadishu before imputation of non-assignment modules (normalized to 100%), and after imputing full consumption

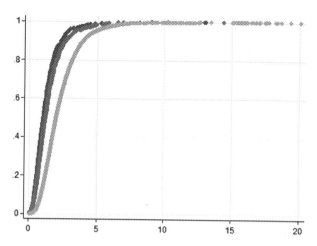

Fig. 4 Cumulative consumption distribution (in USD) per day and per capita (Color figure online) (*Note* For core module (dark blue), core and assigned optional modules (medium blue), and imputed consumption (light blue). The presented consumption aggregate does not include consumption from durable goods

The effect of a response bias due to this can neither be estimated from the simulations nor from the data collected in Mogadishu. However, an enhanced design with different optional modules varying in their comprehensiveness can shed light on this bias. Comparison between responses for the same item in a comprehensive and an incomprehensive list would indicate a lower bound for response bias. Assuming that a comprehensive list results in a better estimate, the response bias could be corrected.

The rapid consumption methodology can increase the gap between capacity at enumerator level and the complexity of the survey instrument. Capacity at the enumerator level is often low in developing countries, especially in a fragile context. The rapid consumption methodology increases the complexity of the questionnaire, which can further increase the gap between existing and required enumerator capacity. However, CAPI technology can seal off complexity from enumerators, as software can automatically create the consumption module based on core and optional modules for each household without showing the partition to the enumerator. In Mogadishu, advanced CAPI technology was used to automatically generate the questionnaire based on the assignment of the household to an optional module. While enumerators were made aware that different households would be asked about different items, administering the rapid consumption questionnaire did not require any additional training of enumerators beyond that needed for a standard consumption questionnaire.

Analysis of rapid consumption data requires high capacity. Analysis capacity is usually limited in developing countries, and especially in fragile contexts. While the general idea of optional consumption modules being assigned to households is digestible by local counterparts, poverty analysis based on a bootstrapped sample of consumption distribution is likely to overwhelm local capacity. However, even standard poverty analysis is often beyond the limits of local capacity in fragile countries. Therefore, capacity building usually focuses on data collection skills with a longer-term perspective on increasing data analysis capacity. In addition, the rapid consumption methodology might be the only way of creating poverty estimates in certain areas, for example, in Mogadishu.

The results of the ex post simulation and the application of the methodology in Mogadishu suggest that the rapid consumption methodology is a promising approach to estimating consumption and poverty in a cost-efficient and fast manner, even in fragile areas.[9] A similar ex post simulation for South Sudan and Kenya (data not shown) indicates that the rapid consumption methodology can also be applied at the country-level, with large intra-country consumption variation.[10] The rapid consumption methodology has been implemented in Somalia, South Sudan, and Kenya, with additional countries in the pipeline.

---

[9]Costs for implementing a rapid consumption survey are lower than conducting a full consumption survey due to the reduced face-to-face time needed, allowing enumerators to conduct more interviews per day.

[10]Ongoing fieldwork is currently employing the rapid consumption methodology in South Sudan to update poverty numbers.

Annex

Consumption of non-assigned optional modules can be estimated by different techniques. Three classes, each with two techniques, are presented here, differing in their complexity and theoretical underpinnings. The first class of techniques uses summary statistics such as the average, to impute missing data. The second class is based on multiple univariate regression models. The third class uses multiple imputation techniques, taking into account the variation absorbed by the residual term.

Summary Statistics (Mean and Median)

This class of techniques applies a summary statistic on the module-specific consumption data collected and applies the result to the missing modules. Each household is assigned the same consumption per missing module. Here, the mean and the median are used as summary statistics. The median has the advantage of being more robust against outliers but cannot capture small module-specific consumption if more than half of the households have zero consumption for the module.

Module-Wise Regression (Ols and Tobit Regression)

Module-wise estimation applies a separate regression model for each module. This allows for differences in core consumption to be captured, as well as other household characteristics. Coefficients are estimated based only on the subsample assigned to the module under consideration. In general, a bootstrapping approach using the residual distribution could mimic multiple imputations, but this is not applied here. Given the impossibility of negative consumption, a Tobit regression with a lower bound of zero is used in addition to a standard OLS regression approach. For the OLS regression, negative imputed values are set to zero.

Multiple Imputation Chained Equations (Mice)

Multiple Imputation Chained Equations (MICE) uses a regression model for each variable and allow missing values in the dependent and independent variables. As missing values are allowed in the independent variables, the consumption of all optional modules can be used as explanatory variables. As a first step, missing values in the explanatory variables are drawn randomly. These values are substituted iteratively with imputed values drawn from the posterior distribution estimated from the regression. While the technique of chained equations cannot be theoretically shown to converge in distribution, the results in practice are encouraging, and the method is widely used.

Multivariate Normal Regression (MImvn)

Multiple Imputation Multivariate Normal Regression uses an expectation-maximization (EM)-like algorithm to iteratively estimate model parameters and missing data. In contrast to chained equations, this technique is guaranteed to converge in distribution with the optimal values. An EM algorithm draws missing data from a prior (often non-informative) distribution and runs an OLS to estimate the coefficients. The coefficients are iteratively updated based on reestimation using imputed values for missing data drawn from the posterior distribution of the model. MImvn employs a data-augmentation (DA) algorithm, which is similar to an EM algorithm, but updates parameters in a non-deterministic fashion, unlike the EM algorithm. Thus, coefficients are drawn from the parameter posterior distribution rather than chosen by likelihood maximization. Hence, the iterative process is a Markov chain Monte Carlo (MCMC) method in the parameter space, with convergence with the stationary distribution that averages the missing data. The distribution for the missing data stabilizes at the exact distribution to be drawn from, to retrieve model estimates averaging over the missing value distribution. The DA algorithm usually converges considerably faster than using standard EM algorithms.

Estimation Performance

The performance of the different estimation techniques is compared based on the relative bias (mean of the error distribution) and the relative standard error. We define the relative error as the percentage difference between the estimated consumptionconsumption and the reference consumption (based on the full consumption module). The relative bias is the average of the relative error. The relative standard error is the standard deviation of the relative error. For estimations based on multiple imputations, the error is averaged over all imputations.

Each proposed estimation procedure is run on the random assignments of households to the optional modules. A constraint ensures that each optional module is assigned equally often to a household per enumeration. The relative bias and the relative standard error are reported across all simulations.

The performance measures can be calculated at different levels. At the household level, relative error is the relative difference in household consumption. At the cluster level, relative error is defined as the relative difference of the average reference household consumption and the average estimated household consumption across the households in the cluster. Similarly, the simulation level compares total average consumption for all households.

References

Beegle, Kathleen, Joachim De Weerdt, Jed Friedman, and John Gibson. (2012). "Methods of Household Consumption Measurement Through Surveys: Experimental Results from Tanzania." *Journal of Development Economics* 98 (1): 3–18.

Christiaensen, L., P. Lanjouw, J. Luoto, and D. Stifel. (2011). "Small Area Estimation-Based Prediction Methods to Track Poverty: Validation and Applications." *Journal of Economic Inequality* 10 (2): 267–297.

Douidich, M., A. Ezzrari, R. van der Weide, and P. Verme. (2013). "Estimating Quarterly Poverty Rates Using Labor Force Surveys." World Bank Policy Research Working Paper no. 6466.

The World Bank. (2018). *Poverty and Shared Prosperity 2018: Piecing Together the Poverty Puzzle.* Washington, DC: World Bank.

9 Rapid Consumption Surveys

The opinions expressed in this chapter are those of the author(s) and do not necessarily reflect the views of the International Bank for Reconstruction and Development/The World Bank, its Board of Directors, or the countries they represent.

Open Access This chapter is licensed under the terms of the Creative Commons Attribution 3.0 IGO license (https://creativecommons.org/licenses/by/3.0/igo/), which permits use, sharing, adaptation, distribution and reproduction in any medium or format, as long as you give appropriate credit to the International Bank for Reconstruction and Development/The World Bank, provide a link to the Creative Commons license and indicate if changes were made.

Any dispute related to the use of the works of the International Bank for Reconstruction and Development/The World Bank that cannot be settled amicably shall be submitted to arbitration pursuant to the UNCITRAL rules. The use of the International Bank for Reconstruction and Development/The World Bank's name for any purpose other than for attribution, and the use of the International Bank for Reconstruction and Development/The World Bank's logo, shall be subject to a separate written license agreement between the International Bank for Reconstruction and Development/The World Bank and the user and is not authorized as part of this CC-IGO license. Note that the link provided above includes additional terms and conditions of the license.

The images or other third party material in this chapter are included in the chapter's Creative Commons license, unless indicated otherwise in a credit line to the material. If material is not included in the chapter's Creative Commons license and your intended use is not permitted by statutory regulation or exceeds the permitted use, you will need to obtain permission directly from the copyright holder.

10

Studying Sensitive Topics in Fragile Contexts

Mohammad Isaqzadeh, Saad Gulzar and Jacob Shapiro

1 Motivation

Fragility, conflict, and violence (FCV) drastically undermines the effectiveness and efficiency of providing public goods and services to the poor. FCV is moreover, a difficult field to study because of the sensitivity and complexity of the nature of events to be addressed. To understand how conflict and violence affect development programs and peoples' livelihood in fragile states requires assessing people's perception of the state, insurgent groups, international actors, and actions taken by these actors. Expressing views about these actors and their activities, however, are risky for those living in fragile states. People may fear that expressing their views could cost them potential benefits and that they

---

M. Isaqzadeh (✉) · J. Shapiro
Princeton University, Princeton, NJ, USA
e-mail: mri2@princeton.edu

S. Gulzar
Stanford University, Stanford, CA, USA
e-mail: gulzar@stanford.edu

© International Bank for Reconstruction and Development/The World Bank 2020   **173**
J. Hoogeveen and U. Pape (eds.), *Data Collection in Fragile States*,
https://doi.org/10.1007/978-3-030-25120-8_10

may incur threats by state and non-state actors, stigmatization, and social ostracism. As a result, questions on issues that are perceived to be sensitive can introduce sensitivity bias, that is, respondents may either avoid answering sensitive questions altogether or provide untruthful responses.

Sensitivity biases generally originate from one of four sources: self-image, taboo (intrusive topics), risk of disclosure, and social desirability.[1] Self-image bias refers to untruthful replies based on misperceptions that individuals may have about themselves. Based on self-affirmation theory in psychology, individuals tend to maintain a perception of global integrity and moral adequacy and will reinterpret their own experience until their self-image is restored.[2] Individuals may therefore provide untruthful answers to questions that relate to their integrity and morality because of their distorted self-image, rather than admit an intent to deceive others. The second source of sensitivity bias is taboo or intrusive topics that respondents do not feel comfortable discussing with others. In such cases, non-response is more likely than untruthful answers as individuals try to avoid discussing the topic.[3] Risk of disclosure is the third source of sensitivity bias. Here, respondents are reluctant to reply altogether or provide a truthful response fearing that their response could be disclosed to the government, rebel groups, criminal groups, or local power holders.[4] Risk of disclosure, in the form of security threats by state and non-state actors or social sanctions by the community, is particularly relevant for research in an FCV context

---

[1]Our formulation here and in Sect. 2 draws heavily on Graeme Blair, Alexander Coppock, and Margaret Moor (2018), "When to Worry About Sensitivity Bias: Evidence from 500 List Experiments." Draft. The authors conduct a thorough meta-analysis of more than 500 list experiments (technique explained below).

[2]Steele, Claude M., Steven J. Spencer, and Michael Lynch (1993), "Self-Image Resilience and Dissonance: The Role of Affirmational Resources," *Journal of Personality and Social Psychology* 64 (6): 885–896; Liu, T. J., and G. M. Steele (1986), "Attribution as Self-Affirmation," *Journal of Personality and Social Psychology* 51: 351–340.

[3]Tourangeau, Roger, Lance J. Rips, and Kenneth Rasinski (2000), *The Psychology of Survey Response*. Cambridge: Cambridge University Press.

[4]Blair et al. (2018).

10 Studying Sensitive Topics in Fragile Contexts   175

where the expression of views on sensitive topics could be very costly for individuals.[5]

Finally, social scientists have long identified social desirability, the fourth source of bias, as a common threat to the validity of research findings.[6] Social desirability refers to 'the tendency on behalf of the subjects to deny socially undesirable traits and to claim socially desirable ones, and the tendency to say things which place the speaker in a favorable light.'[7] Social desirability usually reflects a respondent's concern about favorable attitudes of a reference group. The reference group could be peers, bystanders, family members or relatives present at the interview or even broader groups such as one's community or other communities, institutions, or individuals that consume the research findings.[8] An important reference group whose presence could introduce social desirability bias includes researchers and surveyors. In this case, social desirability is sometimes referred to as the 'experimenter demand effect.' In a study of anti-American sentiment in Pakistan, social desirability bias (social image) is found to potentially lead to the underestimation or overestimation of attitudes toward sensitive issues depending on whether those with extreme views conform to, and express views consistent with moderate respondents, and vice versa.[9]

Experimenter demand effects highlight that even if a survey or experiment is conducted in a private context where peer pressure is ruled

---

[5]Reminders of local insecurity reduce response rates on sensitive topics more than on other topics in a recent survey experiment in Somalia. Denny, Elaine, and Jesse Driscoll (2018), "Calling Mogadishu: How Reminders of Anarchy Bias Survey Participation," *The Journal of Experimental Political Science*. For an early paper on this challenges of measurement see Bullock, Will, Kosuke Imai, and Jacob N. Shapiro (2011), "Statistical Analysis of Endorsement Experiments: Measuring Support for Militant Groups in Pakistan," *Political Analysis* 19: 363–384.

[6]Nederhof, Anton J. (1985), "Methods of Coping with Social Desirability Bias: A Review," *European Journal of Social Psychology* 15: 263–280; Rosenthal, Robert (1963), "On the Social Psychology of the Psychological Experiment: The Experiment's Hypothesis as Unintended Determinant of Experimental Results," *American Scientist* 51: 268–283; and Rosenthal, Robert (1966), *Experimenter Effects in Behavioral Research*. New York: Appleton Century-Crofts.

[7]Nederhof (1985: 264).

[8]Blair et al. (2018) and Tajfel, Henri, and John C. Turner (1979), "An Integrative Theory of Intergroup Conflict," *The Social Psychology of Intergroup Relations* 33 (47): 74.

[9]Bursztyn et al. (2017).

out, the presence of a researcher alone could introduce bias and prevent respondents from expressing honest views and attitudes.[10] In a randomized experiment, it was demonstrated that participants who did not vote in an election were 20 percentage points less likely to answer the door to participate in a survey when they had been previously informed through a flyer about the survey, relative to those who had not received a flyer.[11] The experiment shows the strength of stigma and shame that respondents may feel upon revealing that they did not vote to a surveyor, a stranger whom they may never interact with again.[12]

Social desirability bias may be even stronger in fragile contexts where social stigma could be costlier for individuals and where the association of surveys with aid and development projects could disincentivize truthful responses.

Regardless of the type, sensitivity bias can introduce two problems in surveys: item non-response and untruthful responses conditional on a response. In the case of item non-response, respondents take part in the survey but eschew answering sensitive questions, which is recorded as 'Don't Know' or 'Refused to Answer.' Item-non-response can lead to an underestimation of sensitive attitudes/behaviors and bias estimates of treatment effects when sensitivity is correlated with treatment status.[13] Untruthful reply conditional on a response reflects cases where respondents do not avoid answering questions but provide deceitful replies. Both of these outcomes undermine research findings. Considering the importance of studying sensitive attitudes, researchers have invested in developing approaches to eliminate or reduce sensitivity biases. Below, we discuss these approaches and highlight whether they address item non-response, untruthful reply conditional on response, or both.

---

[10]Rosenthal (1963, 1966).

[11]Dellavigna et al. (2016).

[12]Dellavigna, Stefano, John A. List, Ulrike Malmendier, and Gautam Rao (2016), "Voting to Tell others," *The Review of Economic Studies* 84 (1): 143–181.

[13]For example, when estimating the correlation between receiving aid and support for militant groups one might worry that respondents in pro-militant communities are more reluctant to express support if they have gotten aid because they fear future aid will would be withheld. They therefore avoid the question at higher rates than those in other communities, leading one to erroneously conclude that receiving aid was negatively correlated with support for militants.

2 Approaches

Researchers in the fields of psychology, economics, and political science have developed a range of approaches to studying sensitive attitudes, which can be very useful for conducting research and data collection in fragile contexts. Endorsement experiments, list experiment, and randomized response are the most commonly used techniques developed to mitigate sensitivity bias. Table 1 summarizes the three techniques, as well as direct questioning, with respect to their ability to mitigate different types of sensitivity biases.[14] The three techniques can clearly improve direct questioning by reducing non-response and bias due to risk of disclosure and social desirability. However, they are costly in terms of sample size (because they leverage statistical inference on

Table 1 Survey approaches and addressing sensitivity biases

Approach (method of eliciting honest response)	Survey response challenge				
	Non-response	Risk of disclosure	Taboo/ intrusive topic	Social desirability	Self-image
Direct questions (anonymity/safety through rapport building)	No	No	No	No	No
Endorsement experiment (anonymity/ safety through obfuscation)	Yes	Yes	Maybe	Yes	No
List experiment (anonymity/ safety through aggregation)	Yes	Yes	No	Yes	No
Randomized response (ano-nymity/safety through noise)	Yes	Yes	No	Yes	No

---

[14]We thank Graeme Blair for excellent advice on how to frame these issues.

178   M. Isaqzadeh et al.

the difference between two groups vs. using the mean in one group), require extensive pre-testing, and cannot address bias due to the intrusiveness of the topic (taboos) and self-image. In this section, we review the three approaches, their advantages, and limitations.[15] At the end of the section, we will provide a brief overview of behavioral approaches to address sensitivity biases.

2.1   Endorsement Experiments

Endorsement experiments aim to mitigate non-response and biases due to social desirability and risk of disclosure by obfuscating the object of study. They were first used to study race relations in the US but were later used for studying support for states, international actors, and militant groups.[16]

Since questions about support for the state or insurgent groups in fragile states could pose safety issues for enumerators as well as respondents, answers to direct questions about the state or insurgents may not elicit honest answers and typically face high non-response rates. The endorsement experiments overcome both issues by obfuscating the object of evaluation. When applied to measuring support for particular political actors, endorsement experiments seek respondents' views about particular policies, instead of asking the respondents to express views about particular groups or individuals. Researchers solicit views of actors by dividing respondents at random into treatment and control groups. In the control group, respondents are simply asked whether or not they support a particular policy. In the treatment group, respondents are asked the same questions but are reminded that the policy is endorsed by the groups or individuals who are the subject of the study. This approach is based on extensive research in social psychology, which

---

[15]For statistical software and several papers employing these methods, see Graeme Blair and Kosuke Imai's excellent website: http://sensitivequestions.org.

[16]Sniderman, Paul M., and Thomas Piazza (1993), *The Scar of Race*. Boston: Harvard University Press; Blair, Graeme, C. Christine Fair, Neil Malhotra, and Jacob N. Shapiro (2012), "Poverty and Support for Militant Politics: Evidence from Pakistan," *American Journal of Political Science*.

10 Studying Sensitive Topics in Fragile Contexts   179

show that individuals are more likely to favor policies that are endorsed by individuals from groups whom they like.[17]

As endorsement experiments avoid direct questioning about sensitive topics, respondents feel more comfortable answering questions, reducing non-response rates. Because this method provides a reasonable degree of plausible deniability, respondents are more likely to provide truthful replies, reducing bias due to risk of disclosure and social desirability. This method can potentially mitigate bias due to taboo (intrusive topics) if researchers can phrase questions in such a way that respondents do not feel that intrusive words are being associated with them. It cannot, however, mitigate biases due to self-image because it does not deal with misperceptions that individuals have about themselves.

In a study on support for Islamist militant groups in Pakistan, researchers included questions about support for the polio vaccination, among other policies.[18] The respondents in control group received the following message: 'The World Health Organization recently announced a plan to introduce universal Polio vaccination across Pakistan. How much do you support such a policy?'

The respondents in the treatment group were administered this slightly different statement and question, one which associated the policy with one of four militant groups active in the country at the time: 'The World Health Organization recently announced a plan to introduce universal Polio vaccination across Pakistan. Pakistani militant groups fighting in Kashmir have voiced support for this program. How much do you support such a policy?'[19]

---

[17]Chaiken, S. (1980), "Heuristic Versus Systematic Information Processing and the Use of Source Versus Message Cues in Persuasion," *Journal of Personality and Social Psychology* 39 (5): 752–766; Petty, Richard E., John T. Cacioppo, and David Schumann (1983), "Central and Peripheral Routes to Advertising Effectiveness: The Moderating Role of Involvement," *Journal of Consumer Research* 10 (2): 135–146; and Wood, Wendy, and Carl A. Kallgren (1988), "Communicator Attributes and Persuasion: Recipients' Access to Attitude-Relevant Information in Memory," *Personality and Social Psychology Bulletin* 14 (1): 172–182.

[18]Blair et al. (2012).

[19]Blair et al. (2012).

180   M. Isaqzadeh et al.

Compared to the direct questions about the militant groups in this study, the endorsement experiment questions received much lower non-response rates. For instance, while the non-response rate for direct questions ranged from 22% (questions about Al-Qaeda) to 6% (questions about the Kashmir Tanzeem), the non-response rate for endorsement experiments was much lower, ranging from 7.6 to 0.6%.

In addition to measuring sensitive attitudes, endorsement experiments can be utilized to study sensitive political behaviors as well. One study used an endorsement experiment to study voting 'no' on a personhood referendum in Mississippi.[20] They administered two slightly different primes among the treatment and control group, as in the following box.

Endorsement experiment assessing behavior

Control group	Treatment group
We'd like to get your overall opinion of some people in the news. As I read each name, please say if you have a very favorable, somewhat favorable, somewhat unfavorable, or very unfavourable opinion of each person	We'd like to get your overall opinion of some people in the news. As I read each name, please say if you have a very favorable, somewhat favorable, somewhat unfavorable, or very unfavourable opinion of each person
Phil Bryant, Governor of Mississippi?	Phil Bryant, Governor of Mississippi, who campaigned in favor of the 'Personhood' Initiative on the 2011 Mississippi General Election ballot?
Very favorable	
Somewhat favorable	
Don't know/no opinion	
Somewhat unfavorable	
Very unfavorable	
Refused	

Source Rosenfeld et al. (2015)

By obfuscating the researcher's intention and object of evaluation, endorsement experiments are useful in reducing non-response bias and recovering estimates of sensitive attitudes. Official results from an anti-abortion referendum in Mississippi in 2011 showed that while

---

[20]Rosenfeld, Bryn, Kosuke Imai, and Jacob N. Shapiro (2015), "An Empirical Validation Study of Popular Survey Methodologies for Sensitive Questions," *American Journal of Political Science*, 1–20.

direct questioning significantly underestimated the votes against the referendum (by close to 20% in most counties) and had significant non-response rates, the endorsement experiment and list experiment— discussed below—reduced item non-response and removed approximately half the underestimate of 'no' votes. In contrast, randomized response methods—also discussed below—almost completely recovered the known vote shares.[21]

A number of studies have utilized endorsement experiments to study a range of sensitive topics, particularly support for the state and insurgents in fragile states.[22] A useful resource on this topic is a comprehensive guide for, and illustration of, questioning strategy, regression methods, and analysis tools (including software package in R) for endorsement experiments.[23]

The advantage of an endorsement experiment is that it obscures the object of the evaluation above and beyond concealing the respondent's answer to the sensitive question. The main disadvantage is that a latent variable model is needed to estimate sensitive behavior and attitudes. In addition, the endorsement effect does not have an obvious scale, e.g. it is unclear a priori how a certain percentage change in support for a policy when it is associated with a group vs. not, would indicate supporting the group strongly to opposing it strongly on a standard Likert scale. Its estimates are also statistically inefficient (in the sense of requiring a larger sample to achieve a given confidence interval) compared to the other indirect methods discussed below.[24]

---

[21]Rosenfeld et al. (2015).

[22]See, for example: Lyall, Jason, Graeme Blair, and Kosuke Imai (2013), "Explaining Support for Combatants During Wartime: A Survey Experiment in Afghanistan." *American Political Science Review* 107 (4): 679–705; and Blair, Graeme, Jason Lyall, and Kosuke Imai, (2014), "Comparing and Combining List and Endorsement Experiments: Evidence from Afghanistan," *American Journal of Political Science* 58 (4): 1043–1063.

[23]Bullock et al. (2011), follow-on the work by Bullock et al. (2011). For the relevant software package in R and analysis tools, refer to http://endorse.sensitivequestions.org/.

[24]Rosenfeld et al. (2015).

182   M. Isaqzadeh et al.

2.2   List Experiments

List experiments try to mitigate sensitivity biases by introducing uncertainty through aggregation. This method, also referred to as an 'item count technique' has been extensively used to study racial attitudes and prejudice as well as voter turnout and vote buying.[25]

Similar to the endorsement experiment, the sample is randomly divided into treatment and control groups. Both groups are asked to mention the total number of items on a list that they view as favorable or unfavorable (or number of actions they have taken), without identifying which specific items are favorable or unfavorable. The two groups receive similar lists except that the response options for the treatment group includes one additional item, the sensitive item which is the subject of the study.

As with endorsement experiments, list experiments can be used to study both sensitive attitudes and behavior.[26] A list experiment to study vote buying in Nicaragua found that almost one quarter of voters were offered gifts or services in exchange for votes while only 3% reported such activities when asked directly.[27] The following box shows the control and treatment statements used for assessing vote buying.

A regression analysis technique can be used to analyze list experiment data and recent work illustrates the application of the method

---

[25]Raghavarao, Damaraju, and Walter T. Federer (1979), "Block Total Response as an Alternative to the Randomized Response Method in Surveys," *Journal of the Royal Statistical Society, Series B (Statistical Methodology)* 41 (1): 40–45; Gonzalez-Ocantos, Ezequiel, Chad Kiewiet de Jonge, Carlos Mel´endez, Javier Osorio, and David W. Nickerson (2012), "Vote Buying and Social Desirability Bias: Experimental Evidence from Nicaragua," *American Journal of Political Science* 56: 202–217; Kuklinski, J., M. Cobb, and M. Gilens (1997), "Racial Attitudes and the 'New South,'" *Journal of Politics* 59 (2): 323–349; and Holbrook, A. L., and J. A. Krosnick (2010), "Social Desirability Bias in Voter Turnout Reports: Tests Using the Item Count Technique," *Public Opinion Quarterly* 74 (1): 37–67.

[26]For examples of research using list experiment to study racial attitudes see Kuklinski et al. (1997) and Kuklinski, J., P. Sniderman, K. Knight, T. Piazza, P. Tetlock, G. Lawrence, and B. Mellers (1997), "Racial Prejudice and Attitudes Toward Affirmative Action," *American Journal of Political Science* 41 (2): 402–419.

[27]Gonzalez-Ocantos et al. (2012).

10 Studying Sensitive Topics in Fragile Contexts 183

investigating racial hatred in the US based on the 1991 National Race and Politics Survey.[28] There is also a wide range of studies that have relied on list experiments for studying sensitive topics.[29]

List Experiment assessing behavior	
Control group	Treatment group
I'm going to hand you a card that mentions various activities, and I would like for you to tell me if they were carried out by candidates or activists during the last electoral campaign. Please, do not tell me which ones, only HOW MANY	I'm going to hand you a card that mentions various activities, and I would like for you to tell me if they were carried out by candidates or activists during the last electoral campaign. Please, do not tell me which ones, only HOW MANY
• they put up campaign posters or signs in your neighborhood/city	• they put up campaign posters or signs in your neighborhood/city
• they visited your home	• they visited your home
• they placed campaign advertisements on television or radio	• they placed campaign advertisements on television or radio
• they threatened you to vote for them	• they threatened you to vote for them
	• they gave you a gift or did you a favor

Source Gonzalez-Ocantos et al. (2012)

The advantage of list experiments is that respondents do not disclose whether the sensitive item applies to them. By concealing which items a respondent has favorable or unfavorable views about, the list experiment can reduce non-response rates and mitigate biases due to the risk of disclosure and social desirability. Since respondents do not actually reveal which items they agree or disagree with, this method could alleviate the respondents' fear of disclosing their views and their concerns about reference groups. By only expressing the number of favorable or unfavorable items, they can deny reference to the sensitive item. This method, however, cannot mitigate biases due to taboo since the intrusive

---

[28]Imai, Kosuke (2011), "Multivariate Regression Analysis for the Item Count Technique," *Journal of the American Statistical Association* 106 (494): 407–417. The software package in R for analysis of list experiments can be obtained at http://list.sensitivequestions.org/.

[29]Blair et al. (2018).

184 M. Isaqzadeh et al.

words need to be mentioned either in the question or options. This method cannot reduce biases due to self-image either. The main drawback of this approach is the problem of floor and ceiling effects. In the example above, if the respondent has experienced all the control items, then an honest response would no longer be obscure as it reveals that the respondent received a gift or favor in exchange for a vote, which is an example of the ceiling effect.[30]

In a comprehensive meta-analysis of list experiments applied to political attitudes and behaviors, the list experiment performs well, both in terms of recovering estimates consistent with direct questions about non-sensitive behaviors and in terms of reducing bias.[31]

2.3 Randomized Response

The randomized response approach is useful for estimating population-level variables by obscuring respondents' truthful answers through introducing noise in the responses.[32] In this approach, respondents rely on a random outcome (such as flipping a coin) to add noise to the response, noise whose distribution the researcher knows, and can thus later remove from population-level summaries of the responses.

Randomized response questions come in two variants. In the disguised response version, the respondent is given two questions (an innocuous question and a sensitive question) and asked to flip a coin or other randomizing device out of sight of the surveyor. The coin flip determines which of the two questions the respondent answers. In the forced response version, the respondent is asked to answer the sensitive question but the randomizing device can determine their answer, obfuscating each individual's answer. The following box provides an illustration of these techniques.

---

[30]Rosenfeld et al. (2015) and Glynn, Adam N. (2013), "What We Can Learn With Statistical Truth Serum? Design and Analysis of the List Experiment," *Public Opinion Quarterly* 77: 159–172.
[31]Blair et al. (2018).
[32]Warner, Stanley L. (1965), "Randomized Response: A Survey Technique for Eliminating Evasive Answer Bias," *Journal of the American Statistical Association* 60 (309): 63–69.

10 Studying Sensitive Topics in Fragile Contexts

Randomized response	
Disguised response	Forced response
Please flip a coin, but do not tell me what you got. If you receive heads answer question A, otherwise answer question B. Do not tell me what you got, just answer the question based on your coin flip Question A: Did your coin land on heads? Yes/No Question B: Have you ever shoplifted? Yes/No	For this question, I want you to answer yes or no. But I want you to consider the number of your dice throw. If 1 shows on the dice, tell me no. If 6 shows, tell me yes. But if another number, like 2 or 3 or 4 or 5 shows, tell me your own opinion about the question that I will ask you after you throw the dice [TURN AWAY FROM THE RESPONDENT] Now you throw the dice so that I cannot see what comes out. Please do not forget the number that comes out Now, during the height of the conflict in 2007 and 2008, did you know any militants, like a family member, a friend, or someone you talked to on a regular basis? Please, before you answer, take note of the number you rolled on the dice

Source Blair et al. (2015)

Although the randomized response approach has not been used as widely as the endorsement and list experiments because it is slightly harder to explain to respondents, it is an effective method for studying sensitive attitudes and behaviors in contexts where the population is familiar with some randomization device such as the dice.[33] The randomized response technique has been used to study social connections and contacts with members of armed groups in Nigeria, which was not only sensitive but could even pose security threats to the respondents and surveyors if inquired about directly. This method has been used for estimating a range of sensitive behaviors, from application faking to cheating and drug use.[34] In the study on Nigeria, a multivariate regression analysis

---

[33]Blair, Graeme, Kosuke Imai, and Yang-Yang Zhou (2015), "Design and Analysis of the Randomized Response Technique," *Journal of the American Statistical Association* 110 (511): 1304–1319.

[34]Donovan, John J., Stephen A. Dwight, and Gregory M. Hurtz (2009), "An Assessment of the Prevalence, Severity, and Verifiability of Entry-Level Applicant Faking Using the Randomized Response Technique," *Human Performance* 16 (1): 81–106; Scheers, N. J., and C. Mitchell Dayton (1987), "Improved Estimation of Academic Cheating Behavior Using the Randomized Response Technique," *Research in Higher Education* 26 (1): 61–69; Goodstadt, Michael S., and Valerie Gruson (2012), "The Randomized Response Technique: A Test on Drug Use," *Journal of the American Statistical Association* 70 (352): 814–818; and Clark, Stephen J., and Robert A. Desharnais (1998), "Honest Answers to Embarrassing Questions: Detecting Cheating in the Randomized Response Model," *Psychological Methods* 3 (2): 160–168.

technique was used, and researchers provided guidance for power analysis and robust design for randomized response and illustration of applying this technique to their study of contacts with armed groups in Nigeria, in addition to a software package in R for data analysis.[35,36]

Validation studies of the randomized response approach have led to mixed results. A number of validation studies have found that the randomized response method leads to less biased estimates than direct questioning and reduces item non-response, although it is not always better than list experiments and endorsement experiments. In a validation of the Mississippi referendum on the 'Personhood Initiative', the authors found that randomized response outperformed other methods in terms of reducing bias.[37] Compared to the actual referendum results, the bias in the weighted estimate of support for the referendum was only 0.04 in the randomized response while it was 0.236 in the direct question, 0.149 in the list experiment and 0.069 in the endorsement experiment. However, this method was not the best in reducing the non-response rate. Although the non-response rate in the randomized experiment (13%) was lower than the direct question method (20%), it was much higher than the non-response rate on the list experiment (2%) and the endorsement experiment (0.003%).

The main disadvantage of a randomized response approach is that it requires respondents to administer randomization, which can lead to high rates of item non-response and even survey and attrition. Furthermore, using randomizing devices or flipping coins may be culturally inappropriate in some contexts. A number of validation studies report high rates of non-response and less valid estimates for randomized response approach than a list experiment although other studies have found more favorable results and smaller non-response rates.[38]

---

[35]Blair, Graeme, Kosuke Imai, and Yang-Yang Zhou (2015), "Design and Analysis of the Randomized Response Technique," *Journal of the American Statistical Association* 110 (511): 1304–1319.

[36]The software package in R can be obtained at http://rr.sensitivequestions.org/.

[37]Rosenfeld et al. (2015).

[38]For the discussion of advantages and disadvantages of randomized response, see Rosenfeld et al. (2015).

2.4 Behavioral Approaches

Behavioral approaches mitigate sensitivity bias through direct observation of behaviors that reveal preferences without direct inquiry about those preferences. Two common approaches to measuring behavior are dictator games (where the participants are asked to decide whether they want to share money with another participant) or 'offer' experiments where the respondents decide whether or not to accept an amount of money. The strength of these approaches is in their indirect measurement of sensitive attitudes and high degree of obfuscating the objective of the research.

Behavioral approaches have been used in studying a range of attitudes and behaviors, such as discrimination and xenophobia, altruism and prosocial behavior, religious beliefs, and anti-American attitudes.[39] For instance, one study uses financial costs to indirectly study anti-American identity in Pakistan.[40] Study participants were given Pakistani Rupees (Rs.) 100 or 500, when the daily wage of a manual laborer is between Rs. 400 and 500, merely for checking a box to thank the donor. As shown in the box below, in one version of the instrument, the donor was local (the Lahore University of Management Science) while in the second version it was foreign (the US government).

---

[39]Studies of discrimination and xenophobia include Becker, Gary S. (1957), *The Economics of Discrimination*. Chicago: University of Chicago Press; Bursztyn, Leonardo, Georgy Egorov, and Stefano Fiorin (2017), "From Extreme to Mainstream: How Social Norms Unravel," NBER Working Paper No. 23415, May 2017; Rao, Gautam (2013), "Familiarity Does Not Breed Contempt: Diversity, Discrimination and Generosity in Delhi Schools," Working Paper, https:// scholar.harvard.edu/rao/publications/familiarity-does-not-breed-contempt-diversity-discrimination-and-generosity-delhi. For altruism and prosocial behavior, see Anderoni, James (1990), "Impure Altruism and Donations to Public Goods: A Theory of Warm-Glow," *Economic Journal* 100: 464–477; DellaVigna, Stefano, John A. List, and Ulrike Malmendier (2012), "Testing for Altruism and Social Pressure in Charitable Giving," *Quarterly Journal of Economics* 127 (1): 1–56; and Ariely, Dan, Anat Bracha, and Stephan Meier (2009), "Doing Good or Doing Well? Image Motivation and Monetary Incentives in Behaving Prosocially," *American Economic Review* 99 (1): 544–555. For studies using monetary offers to study religiosity, see Augenblick, Ned, Jesse M. Cunha, Ernesto Dal B'o, and Justin M. Rao (2012), "The Economics of Faith: Using an Apocalyptic Prophecy to Elicit Religious Beliefs in the Field," NBER Working Paper No. 18641, December 2012; Condra, Luke N., Mohammad Isaqzadeh, and Sera Linardi (2017), "Clerics and Scriptures: Experimentally Disentangling the Influence of Religious in Afghanistan," *British Journal of Political Science*, 1–19.

[40]Bursztyn et al. (2017).

Behavioral approach: Revealed preference	
Local donor	Foreign donor
You are one of 50% who are taking this survey receiving this offer to receive an additional Rs. 100. Funding for this bonus payment comes from LUMS	You are one of 50% who are taking this survey receiving this offer to receive an additional Rs. 100. Funding for this bonus payment comes from the US government
We can pay you Rs. 100 for completing the survey, but in order to receive the bonus payment you are required to acknowledge receipt of the funds provided by LUMS and thank the funder	We can pay you Rs. 100 for completing the survey, but in order to receive the bonus payment you are required to acknowledge receipt of the funds provided by the US government
Option 1: I gratefully thank LUMS for its generosity and accept the payment from them	Option 1: I gratefully thank the US government for its generosity and accept the payment from them
Option 2: I do not accept the payment	Option 2: I do not accept the payment

Source Bursztyn et al. (2012)

The study in Pakistan found that when participants make decision privately and if the source of the funds is the US government, almost one quarter of them forgo the money, Rs. 100.[41] However, when they expect their decision to be public, a significantly smaller proportion (around 10%) rejects the payment. They conclude that since the participants expect the majority to accept the payment from the US government, a substantial number of them (15%) conform to the majority and accept the payment although they would not in private. When the payment is increased to Rs. 500, the rejection rate falls from 25%, but a significant proportion of the participants (10%) still forgo the payment.

3 Practical Issues

In addition to being useful tools in recovering truthful responses, the indirect methods reviewed in this chapter have a number of practical advantages over direct questioning. First, they help reduce survey staff vulnerability, which might be particularly important in conflict settings. By masking the nature of the question itself, survey staff are more likely

---

[41]Bursztyn et al. (2017).

10 Studying Sensitive Topics in Fragile Contexts   189

to be protected when local authorities do not allow sensitive questions being to be asked, despite legal protection. There is also the added benefit that plausible deniability may protect individuals by not revealing their true response at the individual level in case the survey instruments are compromised. These issues typically do not arise in non-conflict settings but can be particularly important when protecting individual responses is critically important.

Although the indirect methods for studying sensitive topics outperform direct questioning in many settings, they also have limitations. First, the indirect methods add noise to the estimates, which means that for any given level of statistical power, much larger samples are required to measure group-level differences.[42] Although scholars have proposed ways to reduce noise and remedy the problem of large samples in some cases (such as using double lists or negatively correlated items in a list experiment), the requirement of a large sample remains an important drawback of these indirect methods.[43] Second, these methods require much more extensive pre-testing and preparation than direct questions, which would increase the costs (both financial and human resources) for studying the same topics and could affect the research timeline as well. Third, although these methods reduce sensitivity bias, they cannot overcome incentive compatibility issues. These methods may not provide incentives for the respondents to reveal their true views and attitudes even if they are assured that their individual views will not be disclosed. In essence, these methods reduce the cost of expressing views as long as respondents are interested in expressing their views. If the respondents see advantages in concealing their views and attitudes, these methods do not provide them with incentives to express their views. Some of the behavioral approaches overcome this problem by imposing costs on the

---

[42]Blair et al. (2018) show that most prior list experiments have been underpowered and recommend using direct questions for all but the most sensitive questions unless large samples can be obtained.

[43]For discussion of how to address ceiling effect and reduce noise in list experiments see Glynn (2013).

subjects if they do not reveal their preferences, but the three indirect methods do not impose such costs.[44]

The most important lesson learned from the studies that have utilized indirect methods, however, is the significance of pre-testing. Endorsement experiments require finding political issues on which the groups in question would plausibly take a stand for and that all relate to the same latent policy dimension. Properly implementing list experiments requires choosing control items so that floor and ceiling effects are avoided for almost all respondents. And randomized response requires finding a culturally appropriate randomization device and choosing the appropriate type of question. In short, all indirect methods require much more pre-testing of questions and instruments than traditional direct question do in order to ensure that they can recover truthful replies in which researchers are interested.

Given the cultural and contextual diversity of FCV contexts, some of these methods may work in some contexts but not in others. It is very important to select the appropriate method taking into consideration the concerns and context where the research is conducted. Finally, if feasible, researchers should consider validating the findings of indirect methods by comparing them with available census data or social media data whenever available.

References

Blair, Graeme, C. Christine Fair, Neil Malhotra, and Jacob N. Shapiro. (2012). "Poverty and Support for Militant Politics: Evidence from Pakistan." *American Journal of Political Science.*

Bullock, Will, Kosuke Imai, and Jacob N. Shapiro. (2011). "Statistical Analysis of Endorsement Experiments: Measuring Support for Militant Groups in Pakistan." *Political Analysis* 19: 363–384.

---

[44]In Burstyn et al. (2017), for instance, the subjects are imposed costs (forgoing payments from the U.S. government) for expressing anti-American identity. Game theory and "offer" experiments use financial incentives to study altruism.

Fair, C. Christine, Neil Malhotra, and Jacob N. Shapiro. (2013). "Democratic Values and Support for Militant Politics: Evidence from a National Survey of Pakistan." *Journal of Conflict Resolution*, 1–28.

Fair, C. Christine, Rebecca Littman, Neil Malhotra, and Jacob N. Shapiro. (2016). "Relative Poverty, Perceived Violence, and Support for Militant Politics: Evidence from Pakistan." *Political Science Research and Methods*.

Lyall, Jason, Yuki Shiraito, and Kosuke Imai. (2015). "Coethnic Bias and Wartime Informing." *The Journal of Politics* 77 (3): 833–848.

Sniderman, Paul M., and Thomas Piazza. (1993). *The Scar of Race*. Boston: Harvard University Press.

The opinions expressed in this chapter are those of the author(s) and do not necessarily reflect the views of the International Bank for Reconstruction and Development/The World Bank, its Board of Directors, or the countries they represent.

Open Access This chapter is licensed under the terms of the Creative Commons Attribution 3.0 IGO license (https://creativecommons.org/licenses/by/3.0/igo/), which permits use, sharing, adaptation, distribution and reproduction in any medium or format, as long as you give appropriate credit to the International Bank for Reconstruction and Development/The World Bank, provide a link to the Creative Commons license and indicate if changes were made.

Any dispute related to the use of the works of the International Bank for Reconstruction and Development/The World Bank that cannot be settled amicably shall be submitted to arbitration pursuant to the UNCITRAL rules. The use of the International Bank for Reconstruction and Development/The World Bank's name for any purpose other than for attribution, and the use of the International Bank for Reconstruction and Development/The World Bank's logo, shall be subject to a separate written license agreement between the International Bank for Reconstruction and Development/The World Bank and the user and is not authorized as part of this CC-IGO license. Note that the link provided above includes additional terms and conditions of the license.

The images or other third party material in this chapter are included in the chapter's Creative Commons license, unless indicated otherwise in a credit line to the material. If material is not included in the chapter's Creative Commons license and your intended use is not permitted by statutory regulation or exceeds the permitted use, you will need to obtain permission directly from the copyright holder.

11

Eliciting Accurate Consumption Responses from Vulnerable Populations

Lennart Kaplan, Utz Pape and James Walsh

1 The Data Demand and Challenge

Accurate data on the key economic variables affecting people who have been forcibly displaced, such as consumption and assets, is essential to understanding their situation and to developing evidence-based policies to support them. Poor information or data inaccuracies can lead to flawed diagnostics and impact assessments, resulting in inefficient use and a waste of limited resources. In the context of displacement, consumption

---

L. Kaplan
University of Göttingen, Göttingen, Germany

Heidelberg University, Heidelberg, Germany

U. Pape (✉) · J. Walsh
World Bank, Washington, DC, USA
e-mail: upape@worldbank.org

J. Walsh
e-mail: james.walsh@nuffield.ox.ac.uk

© International Bank for Reconstruction and Development/The World Bank 2020
J. Hoogeveen and U. Pape (eds.), *Data Collection in Fragile States*,
https://doi.org/10.1007/978-3-030-25120-8_11

193

data is particularly important because malnutrition is rife and mortality rates are high, and interventions using consumption data are needed to support the immediate basic needs of vulnerable populations.

In previous High Frequency Survey (HFS) survey rounds, approximately 45% of Somali Internally Displaced Persons (IDP) households reported food consumption below subsistence levels, and 80%, below recommended levels. It is no surprise that IDP populations report lower consumption levels. IDPs face significant hardship that hinders their potential for generating adequate livelihoods, such as experiencing the loss of a breadwinner, not having any productive assets, or having fallen victim to violence. Indeed, IDPs have much less control over their own livelihoods, employment opportunities are scarce within camps, and a large part of their consumption is provided for through aid by NGOs and international organizations.

Yet, there are also reasons that indicate that the low levels of consumption might be due, at least in part, to misreporting. First, very low levels of consumption are associated with high rates of mortality due to starvation. The observed mortality rates among IDPs, however, does not indicate that mortality increased due to starvation across the country at such a scale.[1] Second, non-IDP households that are statistically similar on observable characteristics report higher levels of consumption than IDP households. While IDPs and non-IDPs may have different opportunities to generate income, it is unlikely that IDPs do not smooth their resources to balance food and non-food consumption in a way that endangers their life. The vulnerability of the population increases the stakes for getting the data right: for policymakers designing programs to support IDPs, spurious data is either unusable or biased.

The potential for surveys to generate information that is systematically biased is well documented. A large body of research focuses on improving the accuracy of self-reported information collected in household surveys.[2] In the context of IDPs, that respondents feel compelled

---

[1]Although data from the USAID led Famine Early Warning Systems Network (FEWS NET) suggest high level of malnutrition, evidence on mortality across the counties is mixed (FEWS NET 2018).

[2]There are a number of mechanisms through which the validity of self-reported information in surveys can be compromised. Some inaccuracies result from cognitive biases—for example, acquiescence or "yea-saying" (Bachman and O'Malley 1984; Hurd 1999), extreme responding

11 Eliciting Accurate Consumption Responses ... 195

to misreport is particularly relevant. Indeed, survey respondents in IDP camps may believe that their responses will influence the provision of humanitarian aid and will thus misreport consumption in an attempt to influence its distribution. If survey respondents are underreporting, the inaccuracies generated in the data are highly problematic. At best, it makes the data spurious and unusable. At worst, it could lead to misallocations of aid, from more vulnerable areas to less vulnerable areas, or from solutions emphasizing sustainability to immediate relief when immediate relief is unnecessary. Given this context, light touch adaptations to the design of the survey that prime the idea of honesty offer to make big improvements to the quality of the data and support provisions the data informs.[3]

2 The Implementation

The experiment included 4145 IDP and 781 non-IDP households across South Sudan in 2017 rolled out in mid to late 2017. To investigate whether consumption might be underreported by IDP populations, households were randomly exposed to a bundle of 'honesty primes.' The treatment had three components, which were simultaneously administered in one treatment arm (Fig. 1). These included an emphasis on the importance of accurate answers at the beginning of the survey, a short fictional scenario, which required passing judgment on the behavior of one of the characters, and additional questions to

---

(Cronbach 1946; Hamilton 1968), and question order bias (Sigelman 1981). Other inaccuracies emerge from conscious but not calculated behavior. Respondents may deliberately misreport information on sensitive subjects not to distort statistics but to maintain their reputation or to abide by political norms (Gilens et al. 1998; Rosenfeld et al. 2016). Some misreporting is purposeful. Individuals may misreport in a calculated fashion to increase earnings in a study context (Mazar et al. 2008) or to shape the results of the study if they believe that it will inform policy. It is not surprising that this problem might arise in the context of development aid, an area rife with perverse incentives (Bräutigam and Knack 2004; Cilliers et al. 2015).

[3]This chapter is a summary of Kaplan, Pape, and Walsh (2018, forthcoming), "Eliciting Accurate Responses to Consumption Questions Among IDPs in South Sudan Using "Honesty Primes", Policy Research Working Paper Series. The World Bank.

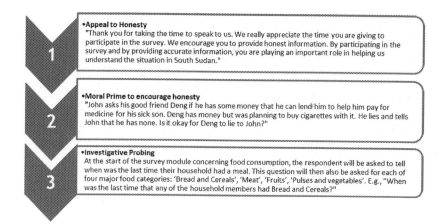

Fig. 1 Treatment Components (*Source* Authors' visualization)

determine the household's last meal, asking respondents to explicitly report whether or not they have eaten in the last week.[4,5] While the former two targets intentional misreporting, the latter addressed classical measurement error.[6] The bundle of primes addressed different psychological mechanisms:

1. Appeals to honesty: These are a standard tool in surveys to increase data accuracy that rely on respondents' preference for the social approval of the enumerator.[7]
2. Honesty primes: These bring the value of honesty to top of mind by asking the respondent to consider a fictional scenario in which honesty is relevant. If individuals feel they have a motivation to misreport, the honesty prime makes a competing motivation salient: to

---

[4]Mazar and Ariely (2006).

[5]One example of this is when individuals' beliefs regarding the consequences of lying affects their behavior. In a two-person experiment where one participant can increase her payoff by lying but at the expense to her counterpart, Gneezy (2005) finds that individuals' propensity to lie is sensitive to the costs it imposes on the other person.

[6]Rasinski et al. (2005) and Vinski and Watter (2012).

[7]Talwar et al. (2015).

answer truthfully to sustain self-consistency. People make decisions on the basis of both external and internal reward systems: even when people have a material incentive to lie, their internal drive to protect their self-integrity may override.[8,9]

3. Investigative probing: This places a higher salience on the importance of getting answer to the question right. By asking for broader categories first, subsequent sub-categories are put under more scrutiny. Self-consistency is reinforced by relating to a longer recall period of seven days.

It is important to note that the treatment is not designed to directly elicit increases in reported consumption. Rather, the intention is to bring the importance of honesty into focus during the interview. It is only through this mechanism—increases in honesty—that we should expect to indirectly see increases in consumption. Thus, ex-ante, we should not expect the treatment effects to be uniform across the consumption distribution.

Almost one-third of respondents (30.1%) reported a calorie intake below the daily subsistence level of 1200 kcal per day and the median per capita consumption was below the recommended calorie intake (1589 kcal per day). Conditioning on adult equivalents, the median shifted well above the recommended daily intake. However, a substantial part of the distribution, 16%, still reported being below the subsistence level and 40% reported being below the recommended daily intake.[10] As with the number of consumption items, the graph indicates that there was a slight shift in the reported consumption among the treated, with respect to very low consumption levels.

---

[8]Mazar and Ariely (2006).

[9]One example of this is when individuals' beliefs regarding the consequences of lying affects their behavior. In an two-person experiment where one participant can increase her payoff by lying but at the expense to her counterpart, Gneezy (2005) finds that individuals' propensity to lie is sensitive to the costs it imposes on the other person.

[10]Several respondents report overly high consumption levels, which far surpass conventional levels (>4000 kcal per day). Robustness checks take this issue into account by censoring the data at the extremes.

Different dependent variables are specified because they have different implications for the respondent's scope of influence on their value. The impact of the 'honesty primes' on the total consumption value, both in terms of money and food intake, is of primary interest. Yet, they are second-order values that are calculated as a function of other variables, including consumption quantities and calories or prices that are in turn deflated. These variables are difficult for respondents to falsify because of the intense mental computation required. The consumption quantity in kilograms is a more direct measure of the quantity consumed as expressed by the respondent and may lead to more accurate estimation of the impact of the 'honesty primes.' Finally, counting the number of items may lead to an even more accurate measure, since the variable is not cleaned and is taken at face value. Furthermore, omitting an item is the easiest and quickest way for respondents to reduce the value of the household's consumption.[11]

3 Key Results

There is a small difference in reported consumption on average between the treatment and control group. The consumption levels shown in Fig. 2 shows a slight difference in consumption between IDP households in the treatment and control groups, though this is apparent only at lower levels of consumption, below SSP 400. In contrast, the distribution of consumption across the two groups matches much more closely for the non-IDP population. The distribution of the number of items displays a similar pattern, though the effect is also faint (Fig. 3). Again, a difference is not visible in the non-IDP population. The number of observations for the non-IDP population is much lower than for the IDP population, and the variance of the distribution is expected to be much greater.

If respondents are deliberately misreporting, those misreporters are likely to be doing so at low consumption levels (e.g., it is more likely to

---

[11]Note that the number of consumption items is not reported at a per-capita level as it does not increase proportionally with household size.

11 Eliciting Accurate Consumption Responses ... 199

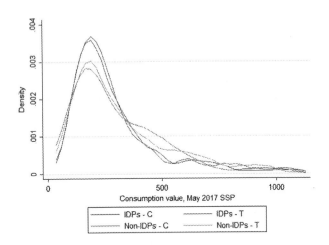

Fig. 2 Consumption distribution by population and treatment (*Source* Authors' calculations using HFS 2017, IDPCSS 2017 and CRS 2017)

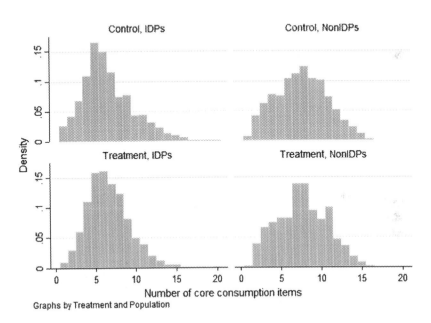

Fig. 3 Number of items consumed by population and treatment (*Source* Authors' calculations using HFS 2017, IDPCSS 2017 and CRS 2017)

be the case that a small number of respondents are significantly under-reporting, rather than a large number of people underreporting by a just a little bit). Given the treatment is not designed to increase reported consumption levels per se, but rather to invoke honesty, it should affect only those people who are misreporting. Hence, heterogenous treatment effects across different household consumption levels (quantiles) test the validity of 'honesty primes.'[12] Figure 4 depicts priming effects across different consumption levels for the four outcome measures of interest.[13] The priming significantly increases reported consumption among lower consumption levels, but not for medium and higher consumption levels. Significant treatment effects mainly influence the reported number of consumption items and the quantities in kilograms. Monetary and caloric consumption measures are not as strongly affected. The latter might also be less susceptible to deliberate misreporting as they depend in part on variables over which the respondent has no control (calories per item; deflators).

The priming has stronger effects among the more vulnerable IDPs. The non-IDP subsample is used to assess the robustness of our main results as we would expect a less significant priming effect among the non-IDPs. Results in Fig. 5 indicate less significant effects, corresponding to the hypothesis that 'honesty primes' are more effective among more vulnerable IDPs.[14] This corresponds to adverse/perverse incentives in foreign assistance settings. Specifically, when IDPs are exposed more intensively to development aid, they may more likely signal their 'neediness' or provide socially desirable answers to signal their 'worthiness' for assistance.

Four dichotomous indicators are used to assess whether the priming shifts a significant share of respondents above certain reporting

---

[12]One might be concerned that honesty primes affect the consumption level of households and, thus, shift the household to another comparison group. Due to the theoretical expectation that treatment effects occur at lower levels of household consumption and are 'light-touch', treatment and control group should still be comparable.

[13]Figure 4 provides a band of the statistical 95% confidence interval of the estimate. Thus, if the confidence band does not cross zero, there would be a 5% chance of indicating significant effects, while the 'true' effect would be zero.

[14]For example, Cilliers et al. (2015) or Bräutigam and Knack (2004).

11 Eliciting Accurate Consumption Responses ...

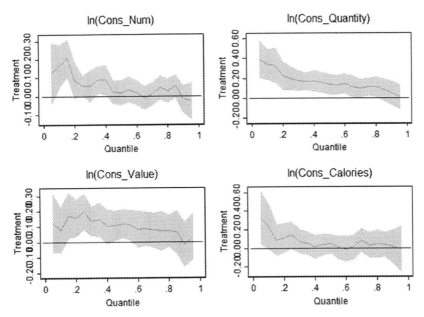

Fig. 4 Treatment effects across quintiles (IDPs) (*Source* Authors' calculations using HFS 2017, IDPCSS 2017 and CRS 2017. All regressions use clustered robust standard errors [White 1980]. Confidence bands refer to the 95% confidence interval. Consumption quantities, values, and calories are used in per-adult equivalent terms. The regression framework is introduced in the appendix. No sampling weights are used as 'honesty primes' are expected to affect, specifically, the extremes of the distribution and the average treatment effect is not a priori of interest)

thresholds. The indicators are equal to one if (i) the respondent household surpasses the caloric subsistence level of 1200 kcal or (ii) the recommended level of caloric intake of 2100 kcal. Two further dummies are created at (iii) 66.66% and (iv) 100% of a normalized poverty line, which is scaled by the fact that only core consumption items were assessed consistently across all surveys. Although the coefficients are mostly positive, only two coefficients turn significant in columns (2) and (3) (Table 1). The results stress the positive effect of the primes, where seven percent more respondent households would have reported above the recommended daily calorie intake level. However, only certain population strata are affected.

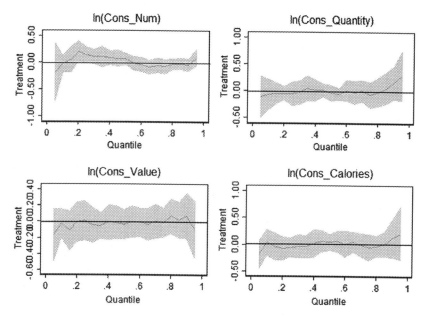

Fig. 5 Treatment effects across quintiles (non-IDPs) (*Source* Authors' calculations using HFS 2017, IDPCSS 2017 and CRS 2017. All regressions use clustered robust standard errors [White 1980]. Confidence bands refer to the 95% confidence interval. Consumption quantities, values, and calories are used in per-adult equivalent terms. The regression framework is introduced in the appendix. No sampling weights are used as 'honesty primes' are expected to affect, specifically, the extremes of the distribution and the average treatment effect is not a priori of interest)

Table 1 Results using poverty thresholds

	(1) >1200 kcal	(2) >2100 kcal	(3) >(2/3) poverty line	(4) >poverty line
Treatment	0.010 (0.027)	0.069* (0.037)	0.063* (0.037)	0.029 (0.036)
Observations	3955	3955	3955	3955
R^2	0.067	0.098	0.118	0.135
State fixed effects	Yes	Yes	Yes	Yes
Month fixed effects	Yes	Yes	Yes	Yes
Controls	Yes	Yes	Yes	Yes
Controls interacted	Yes	Yes	Yes	Yes

Source Authors' calculations using HFS 2017, IDPCSS 2017 and CRS 2017
Robust standard errors in parentheses: *$p<0.1$, **$p<0.05$, ***$p<0.01$

4 Lessons Learned and Next Steps

Most measures to increase the accuracy of surveys assume that respondents want to report as accurately as possible. In many cases, this assumption is incorrect. This research offers novel and suggestive evidence that increasing the salience of honesty may increase survey accuracy, even if incentives to misreport exist. We find significant treatment effects for respondents most likely to be underreporting (those at lower levels), but no significant effects for those at higher levels who are unlikely to be underreporting. We find that the effects are stronger for outcome measures that can easily be manipulated (the number of consumption items) than for those that cannot easily be manipulated (the monetary consumption quantities).

The study underlying this chapter has two main limitations. First, while the experimental set-up allows for identifying a clean treatment effect, it can only compare the control group against an estimate of the 'true' rates of consumption. Without more objective data it is not possible to dismiss the possibility that the higher consumption levels reported in the treatment group are not true and subject to over-reporting. The mortality rates among IDPs suggest that starvation is not occurring systematically across the country, but the precarious situation calls for further scrutiny.[15] Before adjusting poverty estimates, a thorough comparison with more 'objective' data from administrative, anthropometric, or observational sources is needed. Second, the intervention is bundled. For this reason, it is impossible to isolate the causal mechanism affecting the observed changes in reporting. However, if classical measurement error would be affected, treatment effects of the primes should be uniform. In contrast, heterogenous effects across quantiles suggest that the targeting of intentional misreporting via the appeal to honesty and moral prime would be the driver of our results. More research, which unbundles these primes in different treatment arms or combines them with other survey tools can contribute to developing more durable solutions for data collection. Due to both the low

---

[15]FEWS NET (2018).

costs in terms of money and survey time, the 'honesty primes' constitute a valuable supplement for surveys in contexts, where incentives for underreporting exist. Beyond fragile states, the primes could be also a possible survey extension if aid reliance is high (e.g., in Mali or Malawi) as indicated by our subsample analysis.

References

Bachman, J. G., and P. M. O'Malley. (1984). "Yea-Saying, Nay-Saying and Going to Extremes: Black-White Differences in Response Styles." *Public Opinion Quarterly* 48 (2): 491–509.

Bräutigam, D. A., and S. Knack. (2004). "Foreign Aid, Institutions, and Governance in Sub-Saharan Africa." *Economic Development and Cultural Change* 52 (2): 255–285.

Cilliers, J., O. Dube, and B. Siddiqi. (2015). "The White-Man Effect: How Foreigner Presence Affects Behavior in Experiments." *Journal of Economic Behavior and Organization* 118: 397–414.

Cronbach, L. J. (1946). "Response Sets and Validity." *Educational and Psychological Measurement* 6 (4): 672–683.

FEWS NET. (2018). Famine Early Warning Systems Network. Tratto il giorno March 22, 2018 da South Sudan Food Security Outlook—October 2017 to May 2018. http://www.fews.net/sites/default/files/documents/reports/SOUTH%20SUDAN%20Food%20Security%20Outlook_102017_0.pdf.

Gilens, M., P. M. Sniderman, and J. H. Kuklinski. (1998). "Affirmative Action and the Politics of Realignment." *British Journal of Political Science* 28 (1): 159–183.

Gneezy, U. (2005). "Deception: The Role of Consequences." *American Economic Review* 95 (1): 384–394.

Hamilton, D. L. (1968). "Personality Attributes Associated with Extreme Response Style." *Psychological Bulleting* 69 (3): 192–203.

Hurd, M. D. (1999). "Anchoring and Acquiescence Bias in Measuring Assets in Household Surveys." *Journal of Risk and Uncertainty* 19 (1/3): 111–136.

Kaplan, Lennart Christian, Utz Johann Pape, and James Sonam Walsh. (2018, forthcoming). "Eliciting Accurate Responses to Consumption Questions Among IDPs in South Sudan Using 'Honesty Primes'." Policy Research Working Paper Series. The World Bank.

Mazar, N., and D. Ariely. (2006). "Dishonesty in Everyday Life and Its Policy Implications." *Journal of Public Policy & Marketing* 25 (1): 117–126.

Mazar, N., O. Amir, and D. Ariely. (2008). The Dishonesty of Honest People: A Theory of Self-Concept Maintenance. *Journal of Marketing Research* 45 (6): 633–644.

Rasinski, K. A., P. S. Visser, M. Zagatsky, and E. M. Rickett. (2005). "Using Implicit Goal Priming to Improve the Quality of Self-Report Data." *Journal of Experimental Social Psychology* 41 (3): 321–327.

Rosenfeld, B., K. Imai, and J. N. Shapiro. (2016). "An Empirical Validation Study of Popular Survey Methodologies for Sensitive Questions." *American Journal of Political Science* 60 (3): 783–802.

Sigelman, L. (1981). "Question-Order Effects on Presidential Popularity." *Public Opinion Quarterly*, 199–207.

Talwar, V., C. Arruda, and S. Yachison. (2015). "The Effects of Punishment and Appeals for Honesty on Children's Truth-Telling Behavior." *Journal of Experimental Child Psychology*, 130: 209–217.

Vinski, M., and S. Watter. (2012). Priming Honesty Reduces Subjective Bias in Self-Report Measures of Mind Wandering. *Consciousness & Cognition* 21 (1): 451–455.

White, H. (1980). A Heteroskedasticity-Consistent Covariance Matrix Estimator and a Direct Test for Heteroskedasticity. *Econometrica* 48 (4): 817–838.

The opinions expressed in this chapter are those of the author(s) and do not necessarily reflect the views of the International Bank for Reconstruction and Development/The World Bank, its Board of Directors, or the countries they represent.

Open Access This chapter is licensed under the terms of the Creative Commons Attribution 3.0 IGO license (https://creativecommons.org/licenses/by/3.0/igo/), which permits use, sharing, adaptation, distribution and reproduction in any medium or format, as long as you give appropriate credit to the International Bank for Reconstruction and Development/The World Bank, provide a link to the Creative Commons license and indicate if changes were made.

Any dispute related to the use of the works of the International Bank for Reconstruction and Development/The World Bank that cannot be settled amicably shall be submitted to arbitration pursuant to the UNCITRAL rules. The use of the International Bank for Reconstruction and Development/The World Bank's name for any purpose other than for attribution, and the use of the International Bank for Reconstruction and Development/The World Bank's logo, shall be subject to a separate written license agreement between the International Bank for Reconstruction and Development/The World Bank and the user and is not authorized as part of this CC-IGO license. Note that the link provided above includes additional terms and conditions of the license.

The images or other third party material in this chapter are included in the chapter's Creative Commons license, unless indicated otherwise in a credit line to the material. If material is not included in the chapter's Creative Commons license and your intended use is not permitted by statutory regulation or exceeds the permitted use, you will need to obtain permission directly from the copyright holder.

Part III

Other Innovations

12

Using Video Testimonials to Give a Voice to the Poor

Utz Pape

1 The Data Demand and Challenge

South Sudan is a country with a very tumultuous recent history, witnessing more than its share of crises since 2013. The collapse of a fragile peace accord in 2016 led to a renewed military confrontation, while international oil prices simultaneously dropped, depriving South Sudan of its main source of foreign exchange. This triggered a severe fiscal and economic crisis, causing prices to skyrocket, and making many market products unaffordable for the majority of South Sudanese. Thus, securing livelihoods has become more and more difficult, with 66% of the population, a record high, living in poverty. While this number summarizes the country's poverty level, which is important for comparability and analyses to inform policies and programs, the number does not reveal the daily struggles that families face.

U. Pape (✉)
World Bank, Washington, DC, USA
e-mail: upape@worldbank.org

© International Bank for Reconstruction and Development/The World Bank 2020   **209**
J. Hoogeveen and U. Pape (eds.), *Data Collection in Fragile States*,
https://doi.org/10.1007/978-3-030-25120-8_12

The collection of household data is usually a passive process where respondents are asked pre-formulated questions. This constrains the respondents in sharing their own narratives and emphasizing what they feel is important. Giving a voice to the poor beyond being an anonymous and abstract data point is not only helpful to better understand the concerns of the poor, but also to empower them to create a narrative that they own. While some social programs include activities to empower the poor by giving them a voice, the implementation of household surveys is an opportunity that is often missed, in terms of using direct contact with the population across a country to transform a one-sided narrative into one that empowers the poor.

2 The Innovation

To empower the poor and bring humanity to an abstract poverty-related number, we decided to collect short, voluntary video testimonials from people living in South Sudan as part of the High Frequency South Sudan Survey. The High Frequency Survey conducts household interviews in urban and rural areas in South Sudan. The survey is used to collect consumption data in order to estimate poverty, and to measure other socio-economic indicators. As the data is collected using tablets, we decided to utilize the full capability of the tablets by recording voluntary videos after the structured interview if the respondent consented. The video testimonials were subsequently edited, English subtitles were added as translations to the local languages, and noise filters were used to enhance audio quality. The video testimonials were then categorized into themes such as poverty and livelihoods or security and displacement, and were published on the dedicated website www.thepulseofsouthsudan.com.

3 Key Results

The testimonials captured the dire situation in South Sudan, revealing what it is like to live in poverty. They were shown as part of workshops and conferences as well as available on a website. While abstract data may help the government fine-tune its policies, the videos depict the sense of powerlessness, the pain of hunger, and the feelings of hopelessness and disappointment that characterize people's experiences. The testimonials capture the struggle of parents watching their children starve, not being able to provide for them or send them to school, and knowing that tomorrow will not be a better day.

The opportunity for the poor to voice their struggles is a first step toward empowerment, allowing them to share their lives with the world. The testimonials can also serve to inspire policymakers to continue finding innovative ways to help the respondents and millions of others like them to escape poverty. While there is no substitute for quantitative analysis in designing programs and policies, such video testimonials are an effective tool to raise awareness about the concerns of the poorest. They make it clear that poverty is not just a number but a human struggle.

4 Implementation Challenges, Lessons Learned, and Next Steps

We started collecting video testimonials in a pilot, without providing specific training or additional equipment to the enumerators. When we watched these testimonials, we quickly realized that some training was essential. While the videos often started by recording the faces of the respondents, the camera usually moved downwards after a few seconds and ended up recording only their feet or the dust on the ground. Loud wind or other noises sometimes drowned out the voices of the respondents.

To improve the quality of the recordings, we collaborated with journalists and documentary producers to design a one-day training for the enumerators. The training was used to introduce two pieces of very inexpensive but essential equipment: A tripod was necessary to ensure that the camera remained steady and focused on the respondent; and a microphone that could be clipped to the shirt of the respondent ensured that the voice would be audible. The training also included professional guidance on asking open-ended questions to initiate the video testimonial. The success of the training was evidenced in the remarkable quality of video testimonials that were recorded after the training.

During the fieldwork period, there was a decline in the number and the quality of video testimonials. Naturally, enumerators were exposed to various pressures, and were required to conduct as many interviews of sufficient quality as possible. To create more space for video testimonials and to focus on the quality of videos, we introduced monetary incentives for enumerators recording the most and the best video testimonials. The enumerators welcomed this competition with each other, and we saw an increase in the number and the quality of testimonials.

The World Bank's inaugural flagship report, Poverty, and Shared Prosperity 2016: Taking on Inequality, raised concerns about addressing prevalent data gaps in measuring poverty. The World Bank has therefore pledged to ensure that the 78 poorest nations have household-level surveys every three years. To date, 41 of 48 Sub-Saharan African countries have surveys ongoing or planned over the next two years. These surveys also represent an opportunity to give more voice to the poor. Our experience in South Sudan shows that recording testimonials is an extremely low-cost intervention when implemented in conjunction with a household survey. In fact, the additional costs in South Sudan were below US$50k—a small percentage of the overall survey costs. Giving a voice to the poor brings us one step closer to achieving our goals of ending extreme poverty and boosting shared prosperity by 2030.

12 Using Video Testimonials to Give a Voice to the Poor

The opinions expressed in this chapter are those of the author(s) and do not necessarily reflect the views of the International Bank for Reconstruction and Development/The World Bank, its Board of Directors, or the countries they represent.

Open Access This chapter is licensed under the terms of the Creative Commons Attribution 3.0 IGO license (https://creativecommons.org/licenses/by/3.0/igo/), which permits use, sharing, adaptation, distribution and reproduction in any medium or format, as long as you give appropriate credit to the International Bank for Reconstruction and Development/The World Bank, provide a link to the Creative Commons license and indicate if changes were made.

Any dispute related to the use of the works of the International Bank for Reconstruction and Development/The World Bank that cannot be settled amicably shall be submitted to arbitration pursuant to the UNCITRAL rules. The use of the International Bank for Reconstruction and Development/The World Bank's name for any purpose other than for attribution, and the use of theInternational Bank for Reconstruction and Development/The World Bank's logo, shall be subject to a separate written license agreement between the International Bank for Reconstruction and Development/The World Bank and the user and is not authorized as part of this CC-IGO license. Note that the link provided above includes additional terms and conditions of the license.

The images or other third party material in this chapter are included in the chapter's Creative Commons license, unless indicated otherwise in a credit line to the material. If material is not included in the chapter's Creative Commons license and your intended use is not permitted by statutory regulation or exceeds the permitted use, you will need to obtain permission directly from the copyright holder.

13

Iterative Beneficiary Monitoring of Donor Projects

Johannes Hoogeveen and Andre-Marie Taptué

1 Introduction

Mali is a sparsely populated, predominantly desert country with an undiversified economy. It is particularly vulnerable to commodity price fluctuations (gold is a major export), and to the consequences of climate change. Mali has a population of 15 million, 10% of whom are living in the three northern regions of Gao, Kidal, and Timbuktu. High population growth rates, low agricultural productivity, and weather shocks fuel food insecurity, poverty, and instability. The delivery of services within this large territory is challenging and affects geographic equity and social cohesion.

Mali's political and security situation became volatile in 2012 when the northern regions were occupied by rebel and criminal groups who

---

J. Hoogeveen (✉) · A.-M. Taptué
World Bank, Washington, DC, USA
e-mail: jhoogeveen@worldbank.org

A.-M. Taptué
e-mail: ataptue@worldbank.org

© International Bank for Reconstruction and Development/The World Bank 2020 **215**
J. Hoogeveen and U. Pape (eds.), *Data Collection in Fragile States*,
https://doi.org/10.1007/978-3-030-25120-8_13

threatened to take over the country in a coup. These events led to a coup and the deployment of French-led military forces in January 2013. In July 2013, the United Nations Multidimensional Integrated Stabilization Mission in Mali (MINUSMA) took over security measures from the French forces. Constitutional order was restored when two-round presidential elections were held in July and August 2013, with a turnout of 49 and 46% of eligible voters, respectively.

A Peace Accord between the government and two rebel coalitions, known as the "Platform" and "Coordination" groups, was signed by the government and the Platform group on 15 May 2015, and by the government and the Coordination group on 20 June 2015. However, its implementation remains challenging. Security, which is critical to ensuring economic recovery and poverty reduction, remains fragile, with continuing attacks on the UN forces and the Malian army by jihadist groups in the north. There are also attacks on civilians in Bamako, the most recent of which targeted the Radisson Blu Hotel in November 2015, the Nord-Sud Azalai Hotel in March 2016, and a holiday resort near Bamako in June 2017.

Following the presidential elections, a Mali donor conference was organized in Belgium. At the conference, the international community confirmed its continued support, and aid flows, which had declined following the coup, resumed. Following the conference, development partners including the World Bank started to prepare new projects, many focusing on the still insecure northern part of the country. With this refreshed engagement came an increased commitment to project performance.

Information on project implementation is typically captured by project monitoring systems. These monitoring systems track progress but are also expected to flag potential shortcomings or problems. In practice, most monitoring systems do not act as independent rapporteurs, but focus on producing progress indicators for midterm and final reviews. Even this reduced role is not always well-executed and reports often come too late to help projects improve. Supervision missions offer another source of information on project performance, but there is a limit to the information such missions obtain. After all, why show a team of visiting project supervisors an activity that is facing problems?

Less biased information about the effectiveness of projects comes from evaluations by non-project staff. Typically, these take the form of randomized control trials, or large-scale surveys, such as the Service Delivery Indicator (SDI) Surveys, which measure the quality of service delivery in health and education, or Public Expenditure Tracking Surveys (PETS). The challenge of these data-intensive approaches is not their reliability, but that they are expensive and therefore not able to be repeated frequently. Moreover, they are time-consuming and rarely deliver quick results; sometimes, results only become available after the project has finished.

2 The Innovation

For project managers who want to use monitoring data, information obtained through iterative feedback loops is to be preferred over data from infrequent surveys. After all, if the aim is to improve outcomes, it is important not only to establish what a project's problems are, but also to act to address them and to assess whether the action resolved the issue. The idea behind an iterative feedback loop is to allow a project team to learn lessons from a project's shortcomings and improve its performance. Once action has been taken, one must assess whether the identified deficiencies have been resolved. To allow for regular feedback, data collection should be affordable and focused. Reliable, regular, and inexpensive data are the ideal (see also Box 1).

To meet these requirements, a beneficiary feedback system was designed that is light and low-cost, focused on a select set of issues, and implemented by an independent entity with no stake in the outcomes of the project. This approach has been labeled: Iterative Beneficiary Monitoring or IBM. By keeping data collection focused (few research questions and small samples), IBM facilitates timely data analysis and the rapid preparation of reports. By keeping data collection costs down, frequent data collection becomes feasible. The IBM approach reflects a major difference from more typical monitoring systems that collect the bulk of their information at the beginning, in the middle, and at the end of the project. The approach fits within the thinking on adaptive project design as well as complexity, approaches to project design and implementation that stress the importance of context, collecting feedback and demonstrating flexibility in design and implementation.[1]

Box 1 Beneficiary monitoring is not a new concept, but light monitoring is

IBM is not the first time projects systematically seek feedback from beneficiaries during project implementation. A 2002 social development paper presented lessons learned from Beneficiary Assessments that aimed to amplify the voice of the people for whom development is intended. In the report, Beneficiary Assessment is presented as a tool for managers who wish to improve the quality of development operations. The approach, which is rarely used today, has been applied to over 300 projects in 60 countries; it is qualitative, and relies on a combination of direct observation, conversational interviews, and participant observation.

This qualitative approach differs from IBM in important ways. IBM samples tend to be much smaller, its reports shorter, more factual, and produced within weeks of data collection. The cost of the qualitative approach is also much higher. Where IBM costs never more than $5000 per round of data collection, the average cost of qualitative Beneficiary Assessments was $40,000 per round of data collection. For these reasons the qualitative approach is less suited to serve as an iterative feedback loop that is repeated regularly.

Source L. F. Salmen (2002).

How does IBM work in practice? An iterative feedback loop begins with gaining intimate knowledge of a project. This implies discussions with the project manager and those responsible for project implementation (such as the Project Implementation Unit) to establish trust and to identify issues in need of investigation.[2] Project staff are in an excellent position to reflect on the factors that may be hampering successful project implementation.

---

[1]Andrews et al. (2012) and Bowman et al. (2015).

[2]Agreeing to an iterative feedback system at the project design stage is another way to facilitate collaboration between project monitors and project implementers. Nobody questions the need for financial audits, and the same should hold for iterative monitoring. It is difficult to oppose the development of such a system at the design stage, when everyone is working to design a project that delivers the best possible results.

13 Iterative Beneficiary Monitoring of Donor Projects 219

Core project documents need to be read, starting with the Project Appraisal Document (i.e. the document describing the project, its objectives, and modes of implementation). The Implementation Manual is another important document because it describes how the project is expected to operate in practice. It can also be invaluable for identifying sources of information or standards that can be used to assess the project. Supervision reports, aide memoires, and mission reports help to identify issues of potential concern. Project familiarization is time-consuming and, in itself, an iterative process. It is indispensable if an effective approach to data collection is to be designed, and because it builds trust with the project staff, laying the groundwork for follow-up once results have been produced.

Collecting information from beneficiaries and others at the front-line of service provision (such as staff working in schools, clinics, or farmers' organizations) is at the heart of the iterative feedback approach. Their experience with the project is what ultimately matters. IBM thus focuses on obtaining direct feedback from these beneficiaries. Identifying what information to obtain from whom is an important step in the design of a feedback system. For instance, in a project offering meals to students, the perspective of parents and guardians is critical because they can ascertain that children have eaten. Students can give their views on the quantity and quality of the food and how often they receive it. Head-teachers can confirm whether the money to buy the food arrives on time, Parent Teacher Associations can explain whether procedures are being followed, and those who prepare the food are well-placed to report whether the money they receive is sufficient.

It is thus critical that the iterative system is developed in close collaboration with project managers. They need to provide access to project files (including beneficiary databases needed for sampling) and to validate the methodology and instruments for data collection. If this is not carefully done, project managers may eventually contest the validity of the results, and little follow-up can be expected. While the monitoring team will need to collaborate closely with project management, the team will also need to ensure that the identity of respondents and the locations where data are collected are kept confidential. If this is not done, there is a risk that the results will be biased.

It is important to keep the data collection exercise light, and to resist the temptation to collect more information than is strictly necessary. A project manager's capacity is often constrained, and a project team can only handle so many issues at a time. Given that the approach is iterative, new issues can be addressed in subsequent rounds of data collection and not all issues need not be investigated in the first iteration. This gives the project team the option to prioritize what is most critical or most easily addressed. By keeping the data collection exercise light, the design of data collection instruments is relatively straightforward. Nonetheless, validation of the data collection instruments by project management remains an essential step. This includes pre-testing in a real-life setting and discussing the instruments with key project staff to assure that the right issues are captured in an appropriate way (Fig. 1).

The design phase of the iterative approach is typically the most time-consuming phase, and hence, the most resource intensive. Rapport must be built with project staff and analysts need to familiarize themselves with the details of the project and develop, discuss, and test data-collection instruments and approaches. In comparison, data collection itself is relatively inexpensive. The "golden rule" of IBM is that each round of data collection should cost less than $5000. This is an arbitrary number which is kept deliberately small to force IBM designers to focus on key issues and affordable samples. Given this cost structure, the iterative feedback loop differs fundamentally from typical survey exercises, where data collection is the costliest part of the process. Keeping data

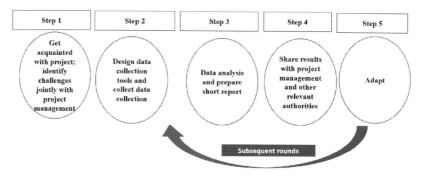

Fig. 1 Five steps of the IBM approach

13 Iterative Beneficiary Monitoring of Donor Projects

collection costs low is of primordial importance to the success of IBM, because in its absence, frequent data collection would not be affordable and its iterative character lost.

Data are typically collected by enumerators specifically hired and trained for the task. Data can be collected using face-to-face interviews, but due to the high transportation costs of survey data collection, samples need to be kept to a minimum. This need not be a problem. When project-related issues are widespread, or when standards or deadlines must be met (as set out in the Implementation or Operations Manual), a small number of deviations may pinpoint a problem. Irrespective of sample size, attention needs to be paid to the sample design to ensure that the results are representative; this implies identifying a database from which the sample can be drawn. This is usually not a problem, as most projects maintain a database of beneficiaries. Additional decisions may also have to be made to keep costs down, but these should always be discussed with project staff, to ensure that such decisions are acceptable. For instance, it may be proposed to sample only from one small geographic area. This might be acceptable, for instance, if this area reflects an upper bound, meaning that the effects of any of the project's shortcomings are likely to be worse in other areas. For example, if it takes a long time to transfer money to schools close to the capital, then it is plausible to assume that the situation is worse in more remote areas.

Figure 2 illustrates a case in Tanzania, generated as a precursor to IBM by one of the authors. It shows how a small number of water kiosks (24 observations), drawn randomly from a database of all water kiosks, already shows that official tariffs set by the regulator are ignored.

Technology can be used to enhance efficiency and reduce cost. If projects collect phone numbers of beneficiaries, information can be collected rapidly and in a cost-effective manner by enumerators who call beneficiaries on their mobile phones (see Chapters 2 and 3 on data collection using mobile phone interviews). This allows for larger samples while remaining within the $5000 data collection budget and is particularly important in a context of insecurity, or when the population may be hostile to authorities and their activities. Mobile phone-based data collection is also a solution when beneficiaries are mobile, as is the case

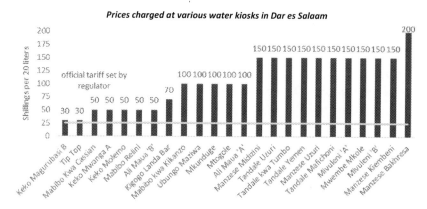

Fig. 2 Small samples may suffice to uncover problems (*Source* Uwazi 2010)

for displaced populations or nomads (Chapter 4). Because collecting data over the phone is inexpensive, collecting phone numbers of beneficiaries simplifies the creation of an iterative feedback loop.

Box 2 How IBM compares to project monitoring

Iterative beneficiary monitoring is an agile, inexpensive way to obtain feedback on project implementation. IBM can be considered a complement to project monitoring in the following ways:

First, while traditional project monitoring is used to continuously assess overall implementation progress and tends to produce voluminous progress reports at fixed points in time, IBM is demand-driven, produces short reports, can be repeated as often as is needed and is focused on diagnosing specific barriers to effective implementation.

Second, project monitoring provides progress reports to the project manager, while IBM reports to the person responsible for the project in the donor organization. IBM thus functions as an independent check on project monitoring systems, much in the same way that financial audits serve as an independent check on companies' regular financial reports. Within the World Bank, IBM is carried out by non-project staff, who do not bear responsibility for supervising the project. Though IBM has never been applied in this manner, it could be viewed as means to assess the ability of an MIS system to identify pertinent issues. By engaging non-project staff, project teams tend to benefit from a fresh perspective that helps teams improve, even in well-established projects.

13 Iterative Beneficiary Monitoring of Donor Projects   223

> Third, relative to a field supervision mission by the project lead, IBM is project supervision "on steroids" as IBM obtains feedback from a much larger sample of beneficiaries than could possibly be covered by a supervision mission visiting two or three project sites. When IBM goes to project sites, it typically visits some 20–30 sites. When beneficiaries are interviewed by phone, sample sizes lie between a couple of hundred and one thousand. IBM also collects data from randomly selected activities, hence avoiding selection bias.

Once collected, data are analyzed and offered as feedback to project managers and project leaders. Given that the dataset is kept small, analysis is rapid. IBM reports are specific, factual and short, and typically less than ten pages. As reports are likely to reveal a project's shortcomings, care is taken to ensure the highest standards of accuracy. Where World Bank projects are concerned, management is copied as a matter of procedure. Often, results will also be discussed with those responsible for the project in the client government. These authorities may request that the project team take the steps required to address the issues but rarely is this needed as project teams tend to be responsive to IBM findings. Another round of data collection will follow sometime later (generally after a few months), with the aim of measuring improvements and, to assess whether new issues may have arisen. The reporting process is the same as for the earlier round. This cycle is repeated on a regular basis until the end of the project.

Reports remain internal, intended for use by the client government, project managers, and supervisors. Disclosing negative facts publicly could have unintended negative consequences, and as is not an objective of IBM.[3] The experience with water price monitoring (as shown in Fig. 2) is illustrative in this regard. Light monitoring principles were applied, but instead of working to address the issue with the regulator, those in charge of the monitoring process sought media attention. Public pressure and parliamentary questions led to corrective action,

---

[3]See also J. Hoogeveen and N. Nguyen (2017).

but these were of an ad hoc and symbolic in nature. Certain responses even aggravated the situation, as some water kiosks were closed because they had been overcharging, leaving those dependent on water kiosks with fewer options than they had previously. After the initial media interest, there was no systematic follow-up, and overcharging continued unabated.

3 Key Results

The IBM approach was first introduced in Mali, offering feedback to an education project (school feeding), an agriculture project (electronic subsidies or e-vouchers), a social protection project (cash transfer), and also to activities managed by the Malian Authorities such as the provision of health insurance to the extreme poor and the functionality of newly established land commissions. In the case of school feeding, the project supervisor expressed concern that only part of the money allocated to this activity was being used. To explore this issue, a clear division of tasks was agreed: the team member from the Poverty Practice of the World Bank would take charge of all issues related to data collection and reporting, while the supervisor from the Education Practice of the Bank would facilitate all interactions with the Ministry of Education and the Project Implementation Unit. The collaboration was smooth, and after some introductory and follow-up meetings, the National Centre of School Canteens at the Ministry of National Education shared the database of schools benefiting from the school feeding program. This database was used to draw a sample of beneficiary schools. To assure ownership and accuracy, officials from the Ministry and the Centre actively participated in the preparation and validation of the survey methodology and tools but were not provided the list of schools included in the sample.

The first round collected data in 20 randomly selected schools. Two enumerators were trained and traveled to each of the schools to carry out face-to-face interviews with head teachers, managers of school canteens, and a subsample of parents. It cost less than US$5000 to complete the data collection exercise, and the report took little time to prepare, as information had only been collected on a limited set of issues. Officials

13 Iterative Beneficiary Monitoring of Donor Projects

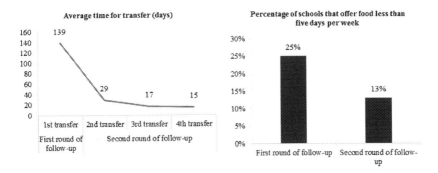

Fig. 3 Regular follow-up improved school feeding performance (*Source* Authors' calculations based on IBM data)

from the National Centre of School Canteens were informed about the main results together with the project manager. Results were shared with the Country Director and the Minister of National Education.

Results showed that it took more than four months to transfer money from the Ministry of National Education to schools. Consequently, much of the money for school feeding arrived after the school year had started, jeopardizing one of the objectives of the program, namely increasing enrolment rates. Moreover, the amount of money sent to schools was insufficient to feed all students during the envisaged period, and some schools were forced to offer food less than five days per week, reducing the incentive for students to remain in school (Fig. 3).

Transfers were expected every quarter, but their real frequency was lower. Also, procedures as described in the operations manual were not followed exactly. Amounts transferred were supposed to reflect enrolment rates for instance, but often they deviated and were much higher or lower than they should have been.

The IBM report was discussed with the project staff, and the Minister of National Education, who followed-up by sending letters to project officials demanding improvements. Additional supervision missions were initiated, and school enrolment information was updated to ensure the correct amounts were transferred.

Six months later, a second round of data was collected, this time in 30 schools randomly selected from a list that excluded the schools that have been interviewed in the first round. Results showed it now took much less time for money to arrive at the schools. Most schools received close to the exact amount that was expected, and all money that was disbursed by the Ministry arrived in the schools. The second report thus showed significant improvements in project implementation, through certain issues persisted (Table 1).

The success of the use of this data collection approach in the education sector aroused interest from other project supervisors. The approach was then introduced to an agriculture project that distributed subsidies in the insecure north of the country using electronic vouchers (e-vouchers). E-voucher beneficiaries had been registered and their phone numbers and core characteristics captured in a database. This information was used to send them vouchers by text message. Upon receipt of their vouchers, beneficiaries could buy specific products, typically fertilizers and livestock products, at designated retail locations at a discount.

Project management expressed concern about the limited uptake of the subsidies. A supervision mission had reported that during the first wave only a fraction of the beneficiaries who had been sent an e-voucher had collected their products, even when they were free of charge. The suggestion was that there might be problems with the distribution system, or that there was a lack of interest among the beneficiaries in the products on offer. Identifying the exact nature of the problems was clearly important for the success of the project.

Because the project had a database with phone numbers of its beneficiaries, and as the areas of intervention were insecure, the team opted to use telephone interviews for data collection. Project management shared its database and participated in working sessions to validate the methodology and survey instruments and to select a representative sample of 100 beneficiaries who were interviewed by phone. Inspection of the shared database revealed the presence of many duplicate phone numbers, allocated to different people in different villages. While the procedural manual permits different beneficiaries to use the same phone number, as not everyone owns a phone, they would be expected to live in the same village. However, the duplicates identified in the database were not in the same location. After four attempts to call a respondent, only 40% had been reached, raising questions about network coverage

Table 1 Iterative feedback approach for a school feeding project in Mali

First round		Actions taken	Second round: six months later
Sample	20 schools	Report discussed with the Minister of National Education	30 schools not included in initial sample
Duration and method for data collection	10 days face to face interviews		10 days, face to face interviews
Cost for data collection	<$5000 US		<$5000 US
Preparation and analysis	5 staff weeks		2 staff weeks
Source of financing	Poverty monitoring task		Poverty monitoring task
Issues	Findings		Findings
1. Time to transfer money to schools	More than three months	Letter by the Minister of Education to project managers	Reduced by 2/3
2. Does the total amount send by central government arrive at schools?	Yes		Yes
3. Does money arrive in a timely manner?	No, arrives late a long time after classes have resumed		Transfer delays have reduced considerably
4. Number of transfers per year	1 out of 4 planned		3 out of 4 planned
5. Number of days covered by amounts sent to school	50% of schools cover less than 40 days as requested	Additional supervision missions	Reduced to 40%
6. Number of days per week food is offered to students	25% of schools offer food less than 5 days per week		Reduced to 13%
7. Do transferred amounts align with enrollment?	Transfers do not reflect for size of the student body, as required		Improved, but a gap persists between school size and numbers used at the Ministry

in villages where beneficiaries live, the accuracy of the phone numbers in the database, and/or the location of beneficiaries, as some people might have left their initial locations due to insecurity.

The initial results showed that all the beneficiaries who had received e-vouchers had collected their products, suggesting that the low uptake of the products was not for a lack of interest. As a significant proportion of beneficiaries could not be reached by phone, it was not possible to know whether all the e-vouchers had been successfully delivered. It seemed plausible that, like the failed telephone interviews, many e-vouchers had failed to reach their intended beneficiaries, suggesting a communication problem between the e-voucher platform and the beneficiaries. Finally, many beneficiaries indicated not having received the full quantity of (free) products indicated on their vouchers. Nor had they been compensated for any items not received.

Following these results, the Bank's team contacted the project and telecom providers to discuss the findings and to address certain issues, including the number of duplicate phone numbers in the database, the inability to send a high number of text messages per second, and the absence of a "text message received" message.

A second round of data collection was carried out five months later. The sample was increased, as there was a need to assess whether the approach was working and how well it worked, as the successful implementation of the e-voucher scheme was a condition for a budget support operation to the government of Mali. More information was needed than a simple understanding of whether the approach was working, and evidence had to be collected on the percentage of beneficiaries in different districts, and the application of targeting criteria. The second round showed that the management of the system had improved. The database was cleaner, more respondents could be reached, more messages could be sent per second, and receipt messages were now received. However, the results also showed that the roll-out of the scheme still left much to be desired. Not all the agreed zones were covered, and e-vouchers had been sent late, three months after the start of the agricultural season. Moreover, e-vouchers were distributed for fertilizers that could not be used given the stage of the growing season. Finally, fertilizer suppliers turned out to have been selected using

Table 2 Iterative feedback for a project distributing agricultural inputs using electronic vouchers

	First round		Second round: 5 months later	
Sample size	100 beneficiaries		850 beneficiaries	
Duration and method for data collection	5 days by phone		10 days by phone	
Cost for data collection	<$5000		<$5000	
Preparation and analysis	3 staff weeks		1 staff week	
Source of financing	Agricultural project		Budget support operation	
Issues	Actions taken		Issues	Actions taken
1. Are beneficiary localities covered by telephone network?	One telecom firm provides information cm its network coverage in localities of the project		1. E-vouchers distributed two months after the start of the planting season	DPO delayed till issues of coverage and timeliness of the e-voucher system are addressed
2. Are vouchers successfully delivered	Meetings organized with telecommunication firms who improved the number of text messages that can be send per second and agreed to send text- receipts. In second round 64% could be reached		2. Only 15% of beneficiaries collected their fertilizer as vouchers had been sent late	
3. Only 40% of beneficiaries can be reached by phone for interview			3. Only 8% of beneficiaries in data base are women	Approach to identifying beneficiaries changed
4. 13% of beneficiaries have duplicates in databases	Database cleaned, duplicates reduced to less than 5%		4. Large price difference between official price of fertilizer and market price (of up to $9 per bag)	Fertilizer now procured using a competitive international procedure
5. 43% of beneficiaries receive less fertilizerthan expected	Measures taken to improve oversight at Hie delivery of inputs; 30% report receiving less than the expected quantities			

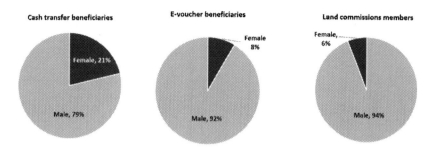

Fig. 4 Selected gender outcomes uncovered by different IBM activities (*Source* Hoogeveen et al. 2018)

a non-competitive method. These findings led to high-level discussions between Work Bank management and the Malian authorities (Table 2).

IBM, because it collects evidence directly from beneficiaries, has proved to be effective at monitoring gender outcomes of projects. In a number of instances, pertinent and concerning gender biases were uncovered. Beneficiaries of a cash transfer program turned out to be mostly men, as were the beneficiaries of the e-voucher program. Land commissions lacked almost any female members (Fig. 4). The adverse gender results uncovered by IBM were not the consequence of bad intentions. Projects were often designed with gender in mind, and in some instances, even employed gender specialists. Invariably, project staff responded positively to the findings when they received them and corrective actions followed. In the latest iteration of IBM, approaches to asking sensitive questions (discussed in Chapter 11) are used to assess from project beneficiaries whether Gender Based Violence might be in issue. Particularly for infrastructure projects in fragile or remote settings this is at times a concern.

4 Implementation Challenges, Lessons Learned, and Next Steps

IBM's iterative feedback approach is relatively straightforward, but applying it successfully requires care. Build a good rapport with a project team is critical, and nobody likes to receive negative feedback,

13 Iterative Beneficiary Monitoring of Donor Projects   231

although this is precisely what an iterative feedback system often does. Confidentiality, good relations with project staff and the government, and agreement on the shared objectives of the monitoring process are essential. Once it is evident that the objectives of the IBM team are aligned with those of the people responsible for project implementation, reticence typically disappears.

Integration of an iterative monitoring approach at the project design stage has the benefit of being able to identify possibilities for beneficiary monitoring early on. Small changes in project design or in the procedural manual can greatly facilitate iterative monitoring. For instance, it makes a difference when procedural manuals stipulate that phone numbers and core characteristics of beneficiaries need to be captured in an electronic database that can be accessed for sampling and (anonymized) monitoring. It also makes a major difference when a procedural manual stipulates that certain benefits need to be distributed by a certain date, as this then offers a clear point in time when progress toward project objectives can be measured.

Even if an iterative monitoring approach is only designed during the project implementation phase, ways can be found to make follow-up monitoring easier. Registering the phone numbers of respondents in face-to-face interviews allows for easy follow-up. Indeed, during each round of the school feeding IBM exercise, phone numbers of respondents (canteen managers, head teachers, and households) were collected for future follow up. Sometimes feedback is offered spontaneously, with beneficiaries volunteering information to the project team, often by text message, about instances when the money for school feeding was exhausted before the expected date, about whether or not the money arrived on time, or about other issues affecting the functioning of the canteen. When such information is received and deemed relevant, the project team can use the phone numbers of other beneficiaries to verify whether what has been reported is a unique case, or an indicator of a more generalized problem.[4]

---

[4]Note that the iterative approach differs from approaches in which beneficiaries are given the opportunity to register complaints. Complaints flag issues, but are not able to distinguish between idiosyncratic negative experiences and the presence of more general project failures. For the latter, feedback needs to be collected in a structured manner.

Another issue for consideration is who should conduct the monitoring. Unsatisfactory results with existing monitoring systems suggest that much is to be said for monitoring by an independent third party. In Mali, staff from the Poverty Practice were responsible for data collection, while staff from the Education respectively Agricultural Practices who were responsible for project implementation, facilitated dialogue with project staff. Working with staff from the Poverty Practice had major advantages. Its micro-economists are experienced in sampling, designing instruments for data collection, training enumerators, and executing primary data collection activities, as well as in data analysis and reporting. Moreover, its staff is familiar with prevailing operating procedures but does not bear responsibility for the success or failure of a project. This facilitates giving independent, unfiltered feedback.

Local presence is another important element for success. Presence facilitates building trust with the project teams and an understanding of how the project operates, and makes it much easier to have discussions about results and corrective actions. Presence close to the location of implementation also increases responsiveness, which is important when issues need to be identified and addressed quickly: after all, lost days cannot be made up, missed meals cannot be replaced, and agricultural inputs distributed late are of little use to farmers.

Familiarity with project procedures and staff facilitates the design of an iterative loop, and as such, outsourcing the approach in the same way as financial audits are outsourced is likely to be a challenge. An intermediate approach, however, could work. Design of instruments and reporting could be left to staff familiar with household survey design and analysis, and dialogue with the client left to those responsible for the project, while data collection could be outsourced. Such an institutional set-up underscores the respective responsibilities of the recipient government, those responsible for project implementation, for project supervision, and for offering beneficiary feedback. It assures a separation of roles which helps avoid reporting bias.

References

Andrews, M., L. Pritchett, and M. Woolcock. (2012). Escaping Capability Traps Through Problem-Driven Iterative Adaptation (PDIA). Faculty Research Working Paper Series RWP12-036, Harvard Kennedy School.

Bowman, C., J. G. Boulton, and P. M. Allen. (2015). *Embracing Complexity. Strategic Perspectives for an Age of Turbulence.* Oxford: Oxford University Press.

Hoogeveen, J., and N. Nguyen. (2017). "Statistics Reform in Africa: Aligning Incentives with Results." *Journal of Development Studies* 55 (4): 702–719.

Hoogeveen, J., D. Kirkwood, A. Savadogo, and A. Taptué. (2018). Enhancing Gender Equality in World Bank Projects Using Iterative Beneficiary Monitoring.

Salmen, L. F. (2002, August). Beneficiary Assessment: An Approach Described. Social Development Papers, Paper number 10. World Bank.

Uwazi. (2010). Water prices in Dar es Salaam. Do water kiosks comply with official tariffs? Policy brief TZ.09/2010E.

The opinions expressed in this chapter are those of the author(s) and do not necessarily reflect the views of the International Bank for Reconstruction and Development/The World Bank, its Board of Directors, or the countries they represent.

Open Access This chapter is licensed under the terms of the Creative Commons Attribution 3.0 IGO license (https://creativecommons.org/licenses/by/3.0/igo/), which permits use, sharing, adaptation, distribution and reproduction in any medium or format, as long as you give appropriate credit to the International Bank for Reconstruction and Development/The World Bank, provide a link to the Creative Commons license and indicate if changes were made.

Any dispute related to the use of the works of the International Bank for Reconstruction and Development/The World Bank that cannot be settled amicably shall be submitted to arbitration pursuant to the UNCITRAL rules. The use of the International Bank for Reconstruction and Development/The World Bank's name for any purpose other than for attribution, and the use of the International Bank for Reconstruction and Development/The World Bank's logo, shall be subject to a separate written license agreement between the International Bank for Reconstruction and Development/The World Bank and the user and is not authorized as part of this CC-IGO license. Note that the link provided above includes additional terms and conditions of the license.

The images or other third party material in this chapter are included in the chapter's Creative Commons license, unless indicated otherwise in a credit line to the material. If material is not included in the chapter's Creative Commons license and your intended use is not permitted by statutory regulation or exceeds the permitted use, you will need to obtain permission directly from the copyright holder.

14

Concluding Remarks: Data Collection in FCV Environments

Johannes Hoogeveen and Utz Pape

Environments characterized by fragility, conflict, and violence (FCV) are very heterogenous, comprising countries as different as Togo and Tuvalu, but also Syria or Chad. Despite this heterogeneity, there are major commonalities. All fragile countries are characterized by limited administrative capacities, country situations are volatile and uncertain, and there is a high degree of data deprivation. Many, but not all, fragile countries are affected by violence. When considering how to address urgent data gaps in fragile countries, the potential for data collectors to get exposed to violence is a defining feature.

In non-violent, fragile countries, efforts should be made to strengthen capacities by rebuilding and strengthening existing statistical systems. As capacities are limited, care should be taken not to overload strained systems with major reform efforts or overly ambitious statistics

---

J. Hoogeveen (✉) · U. Pape
World Bank, Washington, DC, USA
e-mail: jhoogeveen@worldbank.org

U. Pape
e-mail: upape@worldbank.org

© International Bank for Reconstruction and Development/The World Bank 2020
J. Hoogeveen and U. Pape (eds.), *Data Collection in Fragile States*,
https://doi.org/10.1007/978-3-030-25120-8_14

production programs. Certain areas may have to be prioritized, such as the creation of up-to-date sampling frames, as high volatility—often observed in pre- or post-crisis countries—outdates existing sampling frames more rapidly than in a normal context. Given the cost and logistics to update sampling frames with traditional methods, Chapters 7 and 8 offer alternative approaches that could be followed to bridge the gap until a traditional population or enterprise census can take place.

As non-violent, fragile countries are prone to volatility, strengthening the capacity to collect data during times of crisis is recommended. The Rapid Response Surveys discussed in Chapter 3 are especially relevant and could be pursued as part of a more comprehensive crisis readiness approach. The creation of a mobile phone survey team along with the systematic collection of phone numbers of potential respondents, and the preparation of draft phone questionnaires that could be used, are small investments that would yield enormous benefits in terms of information availability during times of distress. Other measures to protect the integrity of the statistical system may also be considered, such as ensuring greater redundancy in the storage of data and reports, including by storing electronic copies off-site or in a cloud.

For fragile countries in which violence is likely, a business as usual approach is neither realistic nor desirable. The monetary as well as opportunity cost of collecting data, whether expressed in financial terms, risk, or use of scarce capacity is much higher in violent settings, and so it is critical to consider whether the envisaged benefits of producing the data are worth the price. Higher cost invariably means less data collection, so trade-offs need to be made. Complex household surveys, suited for non-violent situations, are rarely the instrument of choice in situations of violence. At times complex surveys can be simplified—as is discussed in Chapter 9 using the rapid consumption methodology, but these approaches are technically challenging and for this reason only suited for low capacity environments if complemented with well-trained technical assistance.

When making choices on what to collect, it is important to realize that even in violent situations, many variables remain relatively unchanged over time. Collecting information on such slow-changing aspects should be less of a priority. Other aspects change rapidly in

14 Concluding Remarks: Data Collection in FCV Environments 237

violent FCV environments. Insecurity and deteriorated infrastructure enhance volatility as markets become thinner. As a consequence, food insecurity is more easily at risk. Knowing perceptions, opinions and grievances of citizens is critical as they drive their expectations of the authorities and behavior, including support for local armed groups. So one should first seek answers to questions like: How do prices of key food items change? What is happening to wages, to food security? How are citizen perceptions evolving? How are displaced people cared for? Are interventions succeeding? These aspects should be monitored over time, before moving to more complex surveys.

This suggests that relative to non-violent settings, data collection programs in violent situations should be even more agile. The focus should be on updating information regularly and uncovering trends as opposed to collecting data that gives very precise information about levels. It is more important to know that food security is rapidly worsening than to know what exactly the percentage of food insecure people. This has implications for the way data collection systems are set up. Lighter surveys, or mobile phone surveys should be the standard tools for data collection in FCV settings. Lighter surveys have the advantage that they can be implemented more rapidly, require less capacity for training and analysis. And once call centers have been set up, and phone numbers of different (potential) target groups have been collected, they can be used repeatedly.

With this book we hope to have pointed practitioners to relevant alternatives which can help meet critical data needs, even in the most difficult circumstances. Mobile phone surveys, discussed in Chapters 2–5 give a flavor of the possibilities. When mobile phone surveys are not an option and face-to-face interviews need to be conducted, alternatives can be found by relying on resident enumerators (discussed in Chapter 5), or by designing light data collection instruments like the commune census discussed in Chapter 6. When topics are narrower, for instance whether interventions are succeeding, then iterative beneficiary monitoring (IBM) (Chapter 13) offers an approach that can be followed. When sensitive questions need to be asked or when one is afraid responses might be biased, Chapters 10 and 11 offer pointers.

By sharing these innovations, we hope that many more people can benefit from them, joining us in our attempts to reduce data deprivation and, more importantly, extreme poverty. This book is prepared with practitioners in mind, and when necessary, we focused on showcasing examples rather than elaborating technical details of the approaches. We realize that the proposed approaches vary in complexity, time intensity and cost. Some require a high level of technical expertise at the design stage, others are expensive or difficult to implement. Table 1 may serve as a guide on the kind of expertise that is needed to apply the different approaches discussed in this book.

We welcome feedback and enquiries and are happy to explain in greater depth the methods used and the approaches taken. Contact details for the authors can be found in the section on contributors.

Table 1 Resource requirements to implementing methods described in various chapters

Chapter	Chapter topic	Design complexity	Implementation capacity	Analytical complexity	Cost	Time needed
2	Mobile phone surveys	2	1	2	2	1
3	Rapid reponse survey	3	2	2	2	1
4	Tracking displaced people	3	2	2	2	2
5	Locally recruited enumerators	3	3	2	2	2
6	Local development index	2	2	1	1	1
7	Geo spatial sampling	4	3	3	4	3
8	Sampling displaced populations	3	3	3	2	2
9	Rapid consumption surveys	4	4	4	4	4
10	Studying sensitive topics	3	2	3	1	1
11	Accurate responses	2	1	3	1	2
12	Video testimonials	3	2	1	1	1
13	Iterative beneficiary monitoring	1	1	1	1	1

14 Concluding Remarks: Data Collection in FCV Environments

The opinions expressed in this chapter are those of the author(s) and do not necessarily reflect the views of the International Bank for Reconstruction and Development/The World Bank, its Board of Directors, or the countries they represent.

Open Access This chapter is licensed under the terms of the Creative Commons Attribution 3.0 IGO license (https://creativecommons.org/licenses/by/3.0/igo/), which permits use, sharing, adaptation, distribution and reproduction in any medium or format, as long as you give appropriate credit to the International Bank for Reconstruction and Development/The World Bank, provide a link to the Creative Commons license and indicate if changes were made.

Any dispute related to the use of the works of the International Bank for Reconstruction and Development/The World Bank that cannot be settled amicably shall be submitted to arbitration pursuant to the UNCITRAL rules. The use of the International Bank for Reconstruction and Development/The World Bank's name for any purpose other than for attribution, and the use of the International Bank for Reconstruction and Development/The World Bank's logo, shall be subject to a separate written license agreement between the International Bank for Reconstruction and Development/The World Bank and the user and is not authorized as part of this CC-IGO license. Note that the link provided above includes additional terms and conditions of the license.

The images or other third party material in this chapter are included in the chapter's Creative Commons license, unless indicated otherwise in a credit line to the material. If material is not included in the chapter's Creative Commons license and your intended use is not permitted by statutory regulation or exceeds the permitted use, you will need to obtain permission directly from the copyright holder.

Index

A

adaptive questionnaire
design 46, 47

C

Computer-Assisted Personal
Interviews (CAPI) 10, 70–72,
157, 162, 166
conflict 2, 5, 6, 34, 36, 38,
41, 63, 64, 66, 74, 84–86,
130, 142, 153, 173, 175,
185, 188
consumption 4, 5, 8, 10, 23, 26, 27,
65, 92, 94, 119, 123, 154–
168, 170, 193–195, 197–201,
203, 210, 236
consumption module 7, 10,
157–159, 163, 164, 166,
167, 170

D

displaced 3, 5, 6, 8, 52, 53, 55, 56,
58, 66, 84, 85, 87, 88, 95, 129,
131–133, 142, 193, 222, 237
displaced people 52–58, 60
district census 4, 86–88, 90, 91, 93,
95, 96

E

endorsement experiment 175,
177–182, 186, 190
enumeration area 8, 66, 67, 69, 72,
104–106, 118, 133, 134, 137,
141, 142, 144

F

fragile 2, 3, 5, 6, 10, 153, 154, 166,
167, 173, 176, 178, 181, 204,
209, 216, 230, 235, 236

© International Bank for Reconstruction and Development/The World Bank 2020
J. Hoogeveen and U. Pape (eds.), *Data Collection in Fragile States*,
https://doi.org/10.1007/978-3-030-25120-8

242 Index

fragility 2, 5, 34, 155, 173
Fragility, conflict, and violence
(FCV) 1, 2, 5, 6, 8, 94, 173,
174, 190, 235–237

G

geospatial 142
GIS 112, 116, 124, 125, 137

H

humanitarian 33, 34, 131,
142, 195

I

insecurity 4–6, 37–39, 41, 43, 66,
68, 72, 73, 80, 84, 85, 105,
139, 175, 215, 221, 228, 237
Internally displaced people (IDP) 6,
58, 95, 105, 106, 134, 141,
163, 194, 195, 198
Iterative Beneficiary Monitoring
(IBM) 4, 7, 10, 217–225, 230,
231, 237

L

list experiment 10, 174, 177,
181–186, 189, 190
listing 8, 93, 104, 109–112, 118,
123, 125, 133, 134,
137–141
Local Development Index (LDI) 8,
87, 89, 92, 93, 95–98

M

mobile phone 4, 5, 7, 8, 10, 11,
16–19, 21–24, 36, 45, 47,
48, 54, 60, 61, 67, 87–90, 96,
221, 236, 237
mobile population 67, 113
monitoring 8, 16, 18, 20, 22, 24,
36, 41, 48, 61, 64, 73–77, 81,
84, 87, 92, 153, 216–219,
230–232

P

pastoralist 112
phone interview 4, 17, 19, 22, 55,
67, 221
phone survey 18, 19, 21, 23, 35, 45,
47, 48, 237
plausible deniability 179, 189

Q

qualitative 218

R

randomized response 177, 181, 182,
184–186, 190
Rapid Emergency Response Survey
(RERS) 11, 35, 36, 38
refugee 53, 55, 59, 106, 130–135,
138, 139, 141, 143
resident enumerator 8, 11, 66–68,
77, 237
risk 5, 22, 33–35, 39, 56, 77, 93, 95,
153, 164, 174, 177, 178, 183,
219, 236, 237

Index   243

S

sampling 7, 8, 36, 87, 92, 112, 116, 119, 122, 123, 125, 131–134, 137–141, 147, 201, 202, 219, 231, 232

sampling frame 4–6, 8, 35, 36, 45, 47, 65, 85, 104, 105, 109, 131, 134, 141, 142, 236

survey 5, 6, 8, 10, 11, 17, 19, 20, 22, 23, 34–37, 41–48, 53–55, 57, 58, 60, 65–69, 71–73, 76, 80, 85, 86, 92–95, 106, 109, 112, 113, 116–120, 124, 131, 132, 134, 139, 140, 142, 143, 154, 155, 157, 163, 166, 167, 174–177, 180, 184, 186, 188, 195, 203, 204, 212, 220, 224, 226, 232

T

tablet 11, 72

testimonial 212

tracking survey 59, 61

U

United Nationals High Commissioner for Refugees (UNHCR) 6, 129, 130, 133–135

V

video 8, 10, 11, 18, 210–212

violence 3, 5, 34, 72, 77, 83, 84, 86, 142, 173, 194, 230, 235, 236

W

World Bank 1–3, 16–19, 34, 70, 76, 77, 84, 112, 118, 125, 142, 195, 212, 216, 222, 223

World Food Programme (WFP) 48, 94

Printed in the United States
By Bookmasters